THE

SIERRA NEVADAN WILDLIFE REGION

Fourth Revised Edition

Vinson Brown
and
Barbara Black

Major Illustrators:
Plants: Emily Reid, Janet Coyle, Charles Yocom
Mammals: Jerry Buzzell, Byron Alexander
Birds: Janet Coyle
Reptiles and Amphibians: Robert Stebbins, Byron Alexander

Naturegraph Publishers

Library of Congress Cataloging-in-Publication Data

Brown, Vinson, 1912 - 1991
The Sierra Nevadan Wildlife Region / by Vinson Brown and
Barbara Black; major illustrators, Charles Yocom ... [et al.]. — 4th
rev. ed.
p. cm.
Includes bibliographical references (p.   ) and index.
ISBN 0-87961-227-4 (pbk.)
1. Zoology—Sierra Nevada Region (Calif. and Nev.) 2. Botany-
Sierra Nevada Region (Calif. and Nev.) 3. Animals—Identifica-
tion.
4. Plants—Identification. I. Black, Barbara, 1928-
II. Title.
QH104.5.S54B76 1995
574.9794'4—dc20

95-42041
CIP

ISBN 0-87961-227-4

Books for a better world
Naturegraph Publishers, Inc.
3543 Indian Creek Road
Happy Camp, CA 96039

PREFACE AND APPRECIATIONS

The common and scientific names of the plants and animals in this 1996 revised handbook as well as the species descriptions have been updated using the latest authoritative manuals. For example, for the plants *The Jepson Manual, Higher Plants of California* (Berkeley: University of California Press, 1993) was referred to. The list of suggested references at the end of this book includes most of the sources used for updating the descriptions. In addition, many new plant and wildlife illustrations by qualified nature artists enhance this new edition of *The Sierra Nevadan Wildlife Region*.

I wish to express grateful appreciation to the many people who have assisted in various ways over the years to the revisions and improvement of this book since it first made its appearance in 1954. Dr. Robert Livezey, then in charge of the Sacramento State College Nature School at Lake Tahoe, assisted with the book's first major revision in 1962. Dr. M. R. Brittan, Dr. J.H. Severaid and Dr. H. W. Wiedman, also of Sacramento State College gave advice on the fish, birds, mammals, and plants. For this current revision, Keven Brown, son of Vinson Brown, has been the spearhead. He is nearing completion of his Ph.D. from the University of California at Los Angeles.

It is said that a picture takes the place of many words. Gratitude is extended to the artists whose work complements the descriptions; to McGraw-Hill Book Co. of New York for their kind permission to use the reptile and amphibian pictures of Dr. Robert Stebbins, reproduced from his book on *Amphibians and Reptiles of Western North America*; to the University of California Press at Berkeley, California, for their permission to use drawings of amphibians by Dr. Stebbins, originally reproduced in their book on *Amphibians of Western North America*. Artist initials found in the plant section are ER–Emily Reid, JC–Janet Coyle, CY–Charles Yocom, and KR–Kitty Roach. The mammals are illustrated by Byron Alexander unless initialed JB–Jerry Buzzell. Bird illustrations are mostly Janet Coyle's except for Byron Alexander the hawks, Donald Phillips the Great Gray Owl, and several by JB–Jerry Buzzell. Reptiles and amphibians are a combination of Robert Stebbins' and Byron Alexander's illustrations. For the most part those of the entire animal are by Stebbins and the heads, young, and details by Alexander. The artists for the fish section are BA–Bryon Alexander, FC–Faith Crowder, PT–Phyllis Thompson, and RH–Rune Hapness.

—Barbara Black

iii

TABLE OF CONTENTS

ABOUT THE REGION

Wildlife regions, such as the Sierra Nevadan wildlife region (shown on the map on page 6), are distinctive natural geographic areas of similar climate and topography, which tend to have certain typical animals and plants within their boundaries. There is, however, much overlapping between wildlife regions so that their boundaries should never be considered as rigid lines. The overlap with the Cascade wildlife region is illustrated on the map.

Mountain profile showing the life zones of
THE CENTRAL SIERRA NEVADAN REGION.
Adapted from *TREES, Yearbook of Agriculture,*
1949. U.S. Dept. of Agriculture.

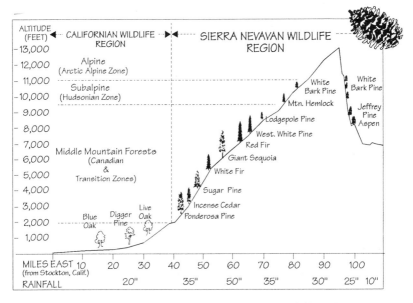

The Transition Zone appears on the west slope with the beginning of the ponderosa pine forest (page 15), where the streamside woodland habitat is also best developed. The Canadian Zone includes the red fir forest (page 16), and parts of the lodgepole pine forest (page 16). These two zones cover the area of the middle mountain forests. The Hudsonian Zone is the same as the sub-alpine forest (page 27), and the Arctic-Alpine Zone is the area of the alpine fell fields and snowy peaks of the mountain tops (page 43). The piñon-juniper woodland is found on the lower east slope of the Sierra Nevada, below the Jeffrey pine forest.

Sierra Nevadan Wildlife Region

Area of overlap with the Cascade Wildlife Region

Major California Mountain Ranges

Klamath

Cascade

North Coast

Sierra Nevada

South Coast

Transverse

Peninsular

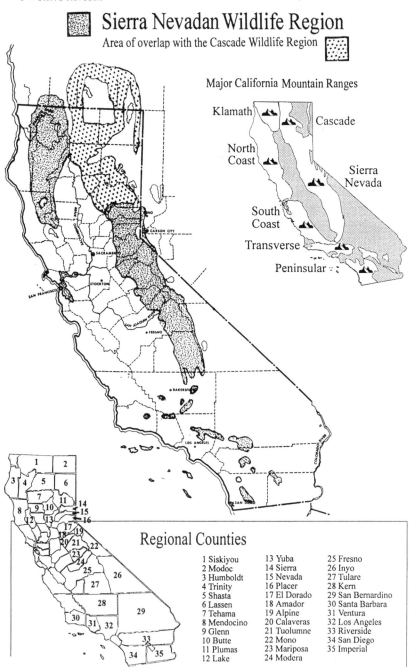

Regional Counties

1 Siskiyou	13 Yuba	25 Fresno
2 Modoc	14 Sierra	26 Inyo
3 Humboldt	15 Nevada	27 Tulare
4 Trinity	16 Placer	28 Kern
5 Shasta	17 El Dorado	29 San Bernardino
6 Lassen	18 Amador	30 Santa Barbara
7 Tehama	19 Alpine	31 Ventura
8 Mendocino	20 Calaveras	32 Los Angeles
9 Glenn	21 Tuolumne	33 Riverside
10 Butte	22 Mono	34 San Diego
11 Plumas	23 Mariposa	35 Imperial
12 Lake	24 Modera	

HOW TO USE THIS BOOK

This book introduces the common wild animals and plants of the Sierra Nevadan wildlife region and provides basic descriptions for their identification. The first step in using this book is to become familiar with the map of the Sierra Nevadan wildlife region on page 6, and with the profile of the mountains showing habitat zones on page 5. Next study the pictures and descriptions of the different habitat sections starting on page 11, so that you can recognize them on a hike or trip. Naturally, there will be some variation in the appearance and terrain of these areas. For example, let us suppose that you are taking a hike and you come into a sub-alpine forest, which you identify from the picture of such a forest on page 27. Beginning with this page, you will find descriptions and illustrations of the common plants of the sub-alpine forest. Look all about you. Study the plants you see. Then, using the descriptions and illustrations, begin your identification. If necessary, first review the section on pages 9 and 10 for helpful information and illustrations on plant identification.

Most plant and animal species live only in certain habitats, which are abbreviated and listed in bold print in the species descriptions. If no geographical range appears with the name of a plant or animal, this means that it is found in most parts of the Sierra Nevadan wildlife region, as shown on the map on page 6. If, however, a range is given, you can tell from this whether the plant or animal is found in your neighborhood. For example, if you live in Weaverville, Trinity County, and find a plant or animal listed as appearing "from Tulare Co. north, but not in the northern Coast Ranges," then you know the plant is not found in your neighborhood, because Weaverville is in the middle of the northern Coast Ranges. If, on the other hand, the description says the plant is "from Tulare Co. north," then you know it is found in your locality because Weaverville is north of Tulare County.

Because plants are often limited to certain wildlife habitats, and thus help to distinguish those areas, all the plant descriptions are grouped under the wildlife habitat sections. For example, all the typical sub-alpine forest plants which include around 90% of the individual plants observed in that habitat, are grouped together under the section on the sub-alpine forest. Some plants that are found more commonly in other habitats, but are also found in the sub-alpine forest, are listed by name at the end of the sub-alpine section, and the page number where each plant is described is given. When a family name is given in a species description, each species following it belongs to the same family until a description occurs with a new family name. Remember, only the common species are described in this book, but these include the majority of the plants you will encounter. For more detailed information on a species, one may refer to the books listed in the Suggested References.

The common animals are listed too at the end of each habitat section. Following each animal's name is a page number where the description of the animal can be found. Because animals are frequently found in many different kinds of wildlife habitats, their descriptions and illustrations are placed together in the last half of the book in the following categories: mammals, birds, reptiles, amphibians, and fishes. Each of these categories opens with a brief introduction giving some information useful for identifying the animals in question.

When walking in a meadow, for example, you suddenly see a bird whose name you would like to know. Turn to the bird section of the book, and look only at the birds that are found in meadows. Another method is to turn to the meadow section of the book and go through the list of birds found there. Then turn to the pages where the descriptions and illustrations of these birds are found and study them until you find the one you believe is right.

ABBREVIATIONS which are used in this book.

{"=inches} {'= feet} {☆=edible plants}

alp.	=	alpine fell fields or meadows
alt.	=	altitude
bldg.	=	buildings
chap.	=	brushland areas
conif.	=	middle mountain forests
FA	=	forearm
fam.	=	family
HB	=	head and body
illus.	=	illustration
L	=	length
mead.	=	middle mountain meadows
pin-jun.	=	piñon-juniper woodlands
rocks	=	rocky areas
sage	=	sagebrush areas
ssp.	=	subspecies
str-wd.	=	streamside woodlands
sub-alp.	=	sub-alpine forests
TL	=	total length
var.	=	variety
W	=	wingspread
water	=	freshwater areas

COMMON PLANTS AND HOW TO IDENTIFY THEM

There are certain successful and widespread plants in each wildlife habitat that help to identify that area. Thus, the willows, dogwoods, and maples predominate in the streamside woodland; the manzanita, deer brush, and whitethorn in the mountain brushland. This book helps you get acquainted with these very common plants, which are the ones you are most likely to see.

The illustrations below show different kinds of flowers, flower formations, and leaves useful in identifying plants. Study these illustrations carefully until you understand the meaning of the plant terms used. Each picture is numbered. When you see one of these numbers in a plant description following a plant term, you can turn to this section to find a helpful illustration of what the term means. When you are studying a plant to determine its name, examine every part of it: the leaves, the flowers, the fruit, the seeds, the bark, and the way the branches are formed. When you add all this knowledge together and compare what you have learned with the plant illustrations and descriptions you are likely to come up with the correct name.

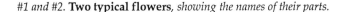

#1 and #2. **Two typical flowers,** *showing the names of their parts.*

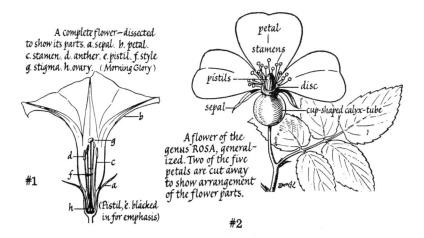

A complete flower—dissected to show its parts. a. sepal. b. petal. c. stamen. d. anther. e. pistil. f. style g. stigma. h. ovary. (Morning Glory)

#1

(Pistil, e. blacked in for emphasis)

petal
stamens
pistils
disc
sepal
cup-shaped calyx-tube

A flower of the genus ROSA, generalized. Two of the five petals are cut away to show arrangement of the flower parts.

#2

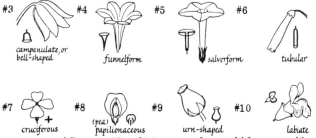

#3 to #10. **Types of flowers:** #3 and #4 are **apetalous**, which means without distinct petals and sepals; #7 and #8 are **choripetalous**, which means the petals and sepals are each completely free from each other; #5, 6, 9 and 10 are **sympetalous**, which means the petals and sepals are all more or less closely joined together.

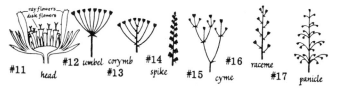

#11 to #17. **Types of flower formations**. The daisy and sunflower look like single flowers, but really are heads of flowers (#11).

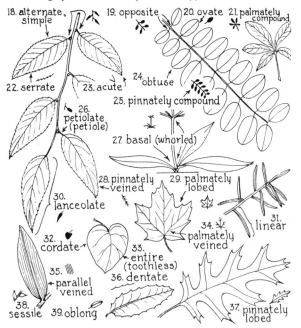

#18 to #39. **Main types of simple and compound leaves.**

THE STREAMSIDE WOODLAND

The streamside woodland is not as well-developed in the Sierra Nevadan region as it is down in the lowlands of the Californian wildlife region. The higher one goes in the mountains the less the streamside woodland is distinct from the surrounding mountain forest, so that in the sub-alpine forest near the mountain tops there is little or no streamside woodland at all. Thus, this habitat is the most developed in the lower part of the middle forest, particularly alongside the ponderosa pine forest (see page 15) where thick growths of willows, dogwoods, maples, and other moisture-loving plants line the streams. This is a favorite habitat of many birds who like a plentiful supply of water and insect life.

1. **SIERRA WATER FERN,** *Thelypteris nevadensis;* Thelypteridaceae, Thelypteris fam. **[str-wd. conif.]** Fronds 1-3′ long, pinnately lobed, with very short stalks (1-2″); lower surface has many resinous glands; forms tufts and grows in large local colonies. From Tuolumne and Trinity counties north, 3000-5000′ altitude.

2. **LADY FERN,** *Athyrium filix-femina;* Dryopteridaceae, Wood Fern fam. **[str-wd. mead. conif. sub-alp.]** Fronds 2–4½′ high, forming large clumps; pinnately-lobed (#37) with the lobe pinnules toothed. From Trinity Co. and southern California mts. north.

3. **FIVE-FINGER FERN,** *Adiantum aleuticum;* Pteridaceae, Brake fam. **[str-wd. mead. rocks, conif.]** Fronds 8–29″ high with distinct five-finger shape; stem reddish brown to blackish. Up to 10,700′ in altitude, from San Gabriel mts. north.

4. **ALPINE LILY,** *Lilium parvum;* Liliaceae, Lily fam. **[str-wd. mead. sub-alp.]** 1–6′ high. Flowers 1–1¼″ long and spotted

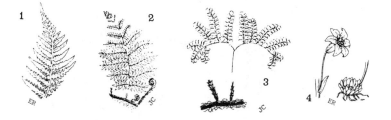

1   2   3   4

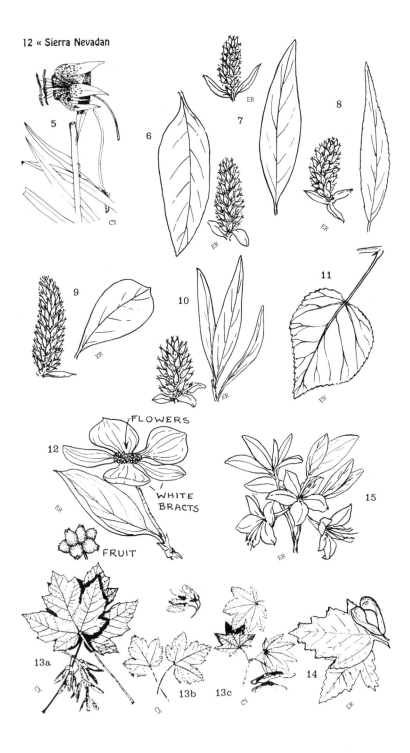

5

CY

6

7

ER

ER

8

ER

9

ER

10

ER

11

ER

12

FLOWERS

WHITE
BRACTS

ER

FRUIT

15

ER

13a

CY

13b

CY

13c

CY

14

ER

purple on orange-yellow. Long, narrow, light green leaves. Usually found above 6000'.

5. **LEOPARD LILY**, *Lilium pardalinum*. **[str-wd. conif.]** Plant 3–7' high. Flowers yellow to orange (tips redder) with maroon spots. Forms large colonies along streams below 6000' altitude.

WILLOWS, (*Salix* species: Salicaceae, Willow and Poplar fam.) Willows are difficult to identify separately, even by experts. 3–30' high; leaves mainly light green in color, linear (#31) to oblong (#39) in shape and with entire (#33) or toothed (#36) edges. The yellow male (S = staminate) and green female (P = pistillate) flowers are in distinctive catkins as is shown. Five common willows are described below.

6. **EASTWOOD'S WILLOW**, or **SIERRA WILLOW**, *Salix eastwoodiae*. **[str-wd. mead.]** A shrub 3–9' high with dark, hairy twigs; stamens gray-hairy. From Tulare Co. north (except in northern Coast Ranges), usually above 6000'.

7. **GRAY-LEAFED SIERRA WILLOW**, *Salix orestera*. **[str-wd. mead.]** 3–6' shrub. Leaves silky and acute at both ends. Not in northern Coast Ranges; usually 7000'+.

8. **LEMMON'S WILLOW**, *Salix lemmonii*. **[str-wd. mead.]** 3–13' shrub. Leaves shiny deep green above; young twigs yellowish; mostly above 6000'.

9. **SCOULER'S WILLOW**, *Salix scouleriana*. **[str-wd. mead. conif.]** 12–30' tree. The commonest willow of middle forest streams; very variable in leaf shape, etc.; young leaves and bark have an unpleasant odor. Below 10,000'.

10. **GEYER'S WILLOW**, *Salix geyeriana*. **[str-wd. mead. conif.]** 3–15' tall shrub or tree. Twigs yellowish to brownish (later darker); very small leaves for a willow.

11. **QUAKING ASPEN**, *Populus tremuloides*. **[str-wd. conif. mead.]** 6–40' tall; graceful and slender tree with smooth, greenish white bark; light green leaves tremble in the breeze. Tulare Co. north. Common in the red fir forest.

12. **MOUNTAIN DOGWOOD**, *Cornus nuttallii*; Cornaceae, Dogwood fam. **[str-wd. conif.]** 12–75' tall tree; characteristic white to pinkish petal-like bracts surround the smaller, center greenish flower heads. The cherrylike fruits are tiny and bright red colored. Green twigs turn dark red to almost black with age.

13A. **BIG LEAF MAPLE**, *Acer macrophyllum*; Aceraceae, Maple fam. **[str-wd. conif.]** Wide-topped tree of 25-90' with palmate (#34) leaves; tiny flowers in racemes (#16); 2-winged seeds. B. **MOUN-TAIN MAPLE**, *A. glabrum.* **[str-wd. conif.]**, has smaller, more compact, tooth-edged leaves, and the seeds have wings at right angles. C. **VINE MAPLE**, *A. circinatum.* **[str-wd. conif.]**. Vine-like, with leaves similar to *A. glabrum*, but even-lobed.

14. **BOX ELDER**, *Acer negundo.* **[str-wd. mead. conif.]** 18–60' tall, round-headed tree. Leaves pinnately 3-lobed, sometimes 5-lobed; ovate and coarsely toothed, densely furry beneath; seeds red, turning yellow.

15. **WESTERN AZALEA**, *Rhododendron occidentale*; Ericaceae, Heath fam. **[str-wd. conif.]** 3–14' high, densely branched shrub with 1–4" long, usually smooth leaves; showy, white or pinkish flowers, 1½–2" long, with one petal usually splotched with yellow; bark shreddy; 1-winged seed.

Other Plants Found in the Streamside Woodland

Western Sword Fern,	17	One-sided Bluegrass,	35	West. Dog Violet,	39
Common Brake Fern,	17	Red Fescue,	35	Nettle-leaved Horsemint,	39
Thimbleberry,	21	Sedges,	35	Indian Paintbrush,	39
Mountain Pink Currant,	21	Corn Lily,	35	Primrose Monkeyflower,	41
Western Bistort,	31	Blue-eyed Grass,	37		

Common Animals of the Streamside Woodland

Mammals

Virginia Opossum,	55	Pacific Jumping Mouse,	69	Western Bluebird,	99
Broad-footed Mole,	55	Norway Rat,	71	Warbling Vireo,	101
Shrew Mole,	55	Mountain Beaver,	71	Solitary Vireo,	101
Vagrant Shrew,	55	Porcupine,	71	Orange-crowned Warbler,	103
Ornate Shrew,	55	Snowshoe Rabbit,	71	Yellow Warbler,	103
Trowbridge Shrew,	55	Mule Deer,	72	MacGillivray's Warbler,	103
Montane Shrew,	55	Elk,	72	Wilson's Warbler,	105
Water Shrew,	55			Brewer's Blackbird,	105
Little Brown Bat,	57	**Birds**		Bullock's Northern Oriole,	105
California Bat,	57	Great Blue Heron,	77	Black-headed Grosbeak,	105
Long-eared Bat,	57	Wood Duck,	79	Lazuli Bunting,	105
Small-footed Bat,	57	Cooper's Hawk,	81	Fox Sparrow,	107
Hairy-winged Bat,	57	Red-tailed Hawk,	81	Purple Finch,	109
Red Bat,	57	Sparrow Hawk,	83	Lesser Goldfinch,	111
Big Brown Bat,	57	Mourning Dove,	83		
Hoary Bat,	57	Killdeer,	85	**Reptiles**	
Silver-haired Bat,	57	Common Snipe,	85	N. Alligator Lizard,	115
Black Bear,	59	Spotted Sandpiper,	85	Rubber Boa,	115
Raccoon,	59	Great Horned Owl,	85	Sharp-tailed Snake,	117
Ring-tailed Cat,	59	Belted Kingfisher,	89	Common Kingsnake,	117
Mink,	59	Northern Flicker,	89	Pacific Ringneck Snake,	117
Long-tailed Weasel,	59	Hairy Woodpecker,	91	Common Garter Snake,	119
River Otter,	61	Downy Woodpecker,	91	Mountain Garter Snake,	119
Striped Skunk,	61	Red-breasted Sapsucker,	91		
Spotted Skunk,	61	Tree Swallow,	93	**Amphibians**	
Gray Fox,	61	Pacific Slope Flycatcher,	91	Rough-skinned Newt,	121
Coyote,	61	Willow Flycatcher,	93	Sierra Newt,	121
Calif. Ground Squirrel,	63	Western Wood Pewee,	93	S. Long-toed Salamander,	121
Western Gray Squirrel,	65	Steller's Jay,	95	Pacific Giant Salamander,	123
Beaver,	67	American Crow,	95	Sierra Nevada Salamander,	123
Deer Mouse,	67	Black-billed Magpie,	95	Black Salamander,	123
Piñon Mouse,	67	Bushtit,	97	Western Toad,	125
W. Harvest Mouse,	67	House Wren,	97	Pacific Treefrog,	125
Dusky-footed Woodrat,	67	American Robin,	99	Foothill Yellow-legged Frog,	125
		Swainson's Thrush,	99	Red-legged Frog,	125
				Northern Leopard Frog,	126

THE MIDDLE MOUNTAIN FORESTS

Typical ponderosa pine forest; photo by U.S. Forest Service.

If you will turn back to the profile of the Sierra Nevada Mountains on page 5, you will see that there is a stretch of thick forest on the mountainside extending from the beginning of the ponderosa pine forest up to the lodgepole forest, between about 2000 and 9000 feet in the central Sierra Nevada. There is a similar, though narrower, stretch of forest on the east side of the mountains. Though this forest includes both the Transition and Canadian life zones and is actually made up of several types of forests, it furnishes a fairly similar habitat for most animal life, and so it is grouped together here under the title "middle mountain forests." Many animals, particularly birds, come into the lower part of this forest in early summer, and gradually move higher as the season gets warmer. Three different main types of forest are found here, each dominated and identified by one kind of tree. Many of the plants that are most common in one of the three types of forest are found less commonly in the others.

a. **Ponderosa Pine Forest** (illustrated at top of this page). This is the lowest coniferous forest of the mountainside, found from 1200–5500 feet in the Mt. Shasta region, 3000–6000 feet in the northern Coast Ranges, 2000–6500 feet in the central Sierra, and 5000–8000 feet in the mountains of southern California. It is a typical Transition Zone forest, fairly open and with little undergrowth. Precipitation runs from 25–80" a year; frost-free days about 90–210; temperature in summer reaches about 80–93 degrees; in winter it drops to 32 degrees or lower. The dominant plants are the ponderosa pine, the white fir, the incense cedar, the sugar pine, and the black oak.

b. **Red Fir Forest.** This is the dense, main forest of the Canadian Life Zone, but the red fir is dominant on the western slope of the Sierra Nevada, whereas the Jeffrey pine is the common tree of this zone on the steeper east slope. The red fir forest appears just above the ponderosa pine forest on the west slope. The precipitation averages 35–65" a year, with very heavy snow; the summer temperature runs about 73–85 degrees at midday, but the winter drops to 26–16 degrees. There are 40–70 frost-free days.

Typical red fir forest; photo courtesy U.S. Forest Service.

c. **Lodgepole Pine Forest.** Trees often close together, but interspersed with frequent meadows; found mainly above the red fir forest as an upper border of the Canadian Life Zone, and sometimes mixed with the lower border of the sub-alpine forest. Precipitation averages about 30–60", mostly snow; frost-free days about 40 or often less; summer temperatures reach 65–75 degrees; winter day time temperatures drop to about 18–10 degrees. Mainly from the central Sierra Nevada north.

Typical lodgepole pine forest; photo courtesy U.S. Forest Service.

FERNS

16. **WESTERN SWORD FERN,** *Polystichum munitum*; Dryopteridaceae, Wood Fern fam. [conif. rocks, str-wd.] 1½-5′ long evergreen fern with the stalks covered with needle-shaped, brown scales; single long fronds appear in clumps; frond-lobes with earlike protuberances at base. Common on wooded hillsides and shaded slopes.

17. **COMMON BRAKE FERN,** *Pteridium aquilinum*; Dennstaedtiaceae, Bracken fam. [conif. mead. sub-alp. str-wd. chap. rocks] Graceful fronds 1-5′ long, erect or ascending, with tiny white hairs beneath; stem stout, generally solitary, with dark cordy rootstocks. Common ground cover, especially under ponderosa pine.

TREES

18. **PONDEROSA PINE,** or **YELLOW PINE,** *Pinus ponderosa*; Pinaceae, Pine fam. [conif.] 100-220′ tall evergreen tree. Short branches turn upward at ends; crown spirelike or sometimes flat-topped in dry localities; needles in 3′s, bright yellow-green, making broomlike tufts on branch ends; bark in old trees forms yellow-brown, picture-puzzlelike plates. Fresh bark has a resinous odor.

19. **JEFFREY PINE,** *Pinus jeffreyi.* [conif. sub-alp.] 90-170′ tall; old bark usually splits into large, reddish brown, irregular plates, but not as picture-puzzlelike as in the ponderosa pine; evergreen needles in 3′s, dull blue-green in color; 4½-6″ long. Young trees are best distinguished from young ponderosa pines by the smell, which is pleasantly vanillalike, and by the cones being larger and the branchlets smoother. Common in dry places, especially on the east side of the Sierra Nevada.

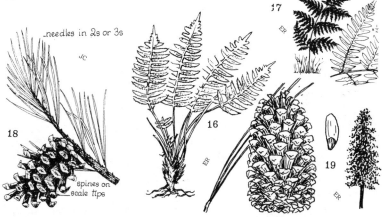

_needles in 2s or 3s

18

spines on scale tips

16

17

19

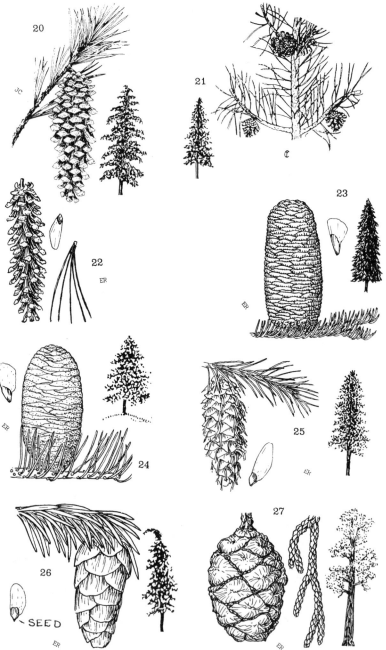

20. ✿ **SUGAR PINE**, *Pinus lambertiana*. [**conif.**] 150-230' tall evergreen tree. The main branches are large and horizontal; young branchlets covered with hair; old bark covered with loose reddish brown scales; needles in 5's, stiff and sharp-pointed; the cones are enormously long; sap sweet, used for sugar by Indians.

21. **LODGEPOLE PINE**, *Pinus contorta murrayana*. [**conif. subalp.**] 60-110' tall evergreen, usually with a very straight trunk except where wind-swept; bark thin and covered with loose scales; slender dark green needles are in 2's; 1-2" long cones are shiny, reddish brown.

22. **WESTERN WHITE PINE**, *Pinus monticola*. [**conif. sub-alp.**] 60-220' high, slender tree with narrow, symmetrical top when in dense forests; more open and loosely-branched in form when found in open forests; mature bark dark gray, relatively thin, and divided into nearly square plates; needles in 5's, usually 2-4" long, bluish green and marked with white lines; cones 4-8" long.

23. **RED FIR**, *Abies magnifica*. [**conif.**] Up to 180' high with short branches, lower ones drooping; young bark gray; mature bark deeply furrowed with dark reddish ridges. The 4-sided, 2-ribbed needles are blue-green, 1-1½" long, and very thick on the branchlets; tree appears spire-shaped.

24. **WHITE FIR**, *Abies concolor*. [**conif.**] An evergreen; may reach 200' high. The short, stout branches usually form a thin, spirelike top; young bark white-gray; old bark gray-brown or darker; the erect, pale blue-green needles are about 2-2½" long and usually grooved; cones 2-4½" long.

25. **DOUGLAS FIR**, *Pseudotsuga menziesii*. [**conif.**] An evergreen, up to 200' high; the lower branches usually droop, and all branches end in many drooping branchlets; old bark with broad ridges and deep grooves. Needles dark green, ¾-1¼" long, and pointing in all directions. From Fresno Co. north.

26. **MOUNTAIN HEMLOCK**, *Tsuga mertensiana*. [**conif. subalp.**] Evergreen; 75-114' high with slender, curved and drooping branches; branches often with up-curved tips; thick reddish brown bark made up of rounded ridges with deep grooves between; needles light blue-green, ½-¾" long; cones 1-3" long. From Tulare Co. north.

27. **GIANT SEQUOIA**, *Sequoiadendron giganteum*; Taxodiaceae, Bald Cypress fam. [**conif.**] An evergreen; up to 288' high and 26' wide at base; old crown irregular, with very large branches; a deeply grooved reddish bark; needles cling to the stems; most of

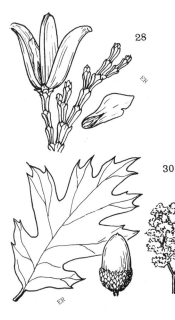

28

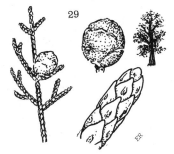

29

30

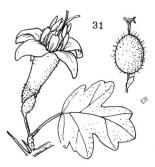

31

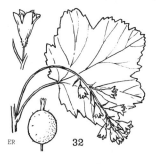

33

32

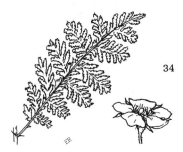

34

35

the reddish brown cones are 1½-3" long. Forms thick groves, from Tulare Co. to Placer Co.

28. **INCENSE CEDAR**, *Calocedrus decurrens*; Cupressaceae, Cypress fam. **[conif.]** An evergreen tree, 65-165' high, with very characteristic light green needles appressed to the branchlets; the bark is bright, reddish brown, and fibrous; cones ¾-1" long.

29.✰ **WESTERN JUNIPER**, *Juniperus occidentalis*. **[conif. pinjun.]** Evergreen; 16-50' high with very thick branches, spreading to form a flat top; brown bark about ⅗" thick; the grayish green needles are opposite or whorled in 3's, and appressed to the branchlets; the round, blue-black berries are edible. East slope of mts.

30.✰ **CALIFORNIA BLACK OAK**, *Quercus kelloggii*; Fagaceae, oak fam. **[conif.]** 30-82' high deciduous tree with smooth bark becoming deeply furrowed with age; leaves are bright green, 4-8" long; acorn reddish brown; usually fine hairy, and edible after the ground meal is leached in warm water. Up to 8000' alt.

SHRUBS

31.✰ **SIERRA GOOSEBERRY**, *Ribes roezlii*; Grossulariaceae, Gooseberry fam. **[conif.]** A 3' high shrub with spreading branches covered with sharp brown spines about ⅓" long; no bristles; leaves ¾" wide, covered with short hairs below; flowers with 5 purplish red sepals, covered with hairs; red berry with glandular spines. Indians mixed berries with other foods for flavoring.

32.✰ **MOUNTAIN PINK CURRANT**, *Ribes nevadense*. **[conif. str-wd. sub-alp.]** 3-7' high shrub with slender, loose branches; old bark forms flakes that drop off; thin, bright green leaves; short, dense racemens (#16) with pink to red flowers; a smooth, blue-black berry, which is edible.

33.✰ **THIMBLEBERRY**, *Rubus parviflorus*; Rosaceae, Rose fam. **[conif. str-wd.]** 3-6' high shrub with smooth branches; large palmate (#29) and cordate (heart-shaped) leaves; a few large, white flowers, which turn into sweet, pale red, raspberry-like berries.

34. **MOUNTAIN MISERY**, *Chamaebatia foliolosa*. **[conif.]** 1-2' high shrub, erect, with many leafy branches; bark smooth, dark brown; ¾-4" long leaves feel sticky; white flowers in cymes (#15); bush smells like creosote and is sometimes used as medicine for colds, etc.

35. **BUSH CHINQUAPIN**, *Castanopsis sempervirens*; Fagaceae, Oak fam. **[conif. rocks, chap.]** 3-8' high, spreading bush with a round top, and smooth, brown bark; 1½-3" long leaves are golden

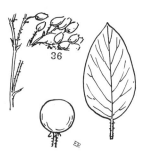

36

37

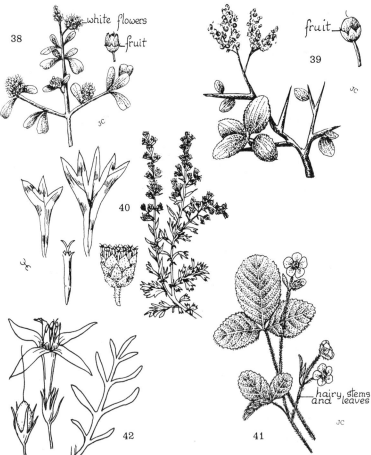

38   white flowers

fruit

fruit

39

40

hairy stems
and leaves

42   41

or rusty colored, fuzzy below, and dull green above; tiny flowers in dense spikes (#14), turning into spiny burrs. Dry places.

36.☆ **MARIPOSA MANZANITA**, *Arctostaphylos mariposa*; Ericaceae, Heath fam. **[conif. chap.]** 5-15' high, erect shrub with a reddish brown smooth bark; branchlets covered with tiny hairs; leaves very pale gray-green; pale pinkish or white, urn-shaped (#9) flowers; cherrylike fruit is red-brown and surrounded by dry, triangular bracts; has acidic taste, but was sometimes used by Indians to mix with other foods.

37.☆ **GREEN-LEAVED MANZANITA**, *Arctostaphylos patula*. **[conif. chap.]** 3-6½' high shrub with very crooked, stiff branches; bark smooth and bright red brown; leaves bright green; pinkish flowers in large panicles (#17); fruit reddish brown and acidic; edible.

38. **DEER BRUSH**, *Ceanothus integerrimus*; Rhamnaceae, Buckthorn fam. **[conif. chap.]** 3-13' high shrub with diffuse branches, covered with pale green bark; the light green leaves usually entire (#33) and smooth; end branches not spiny; sweet-smelling flowers usually white or blue.

39. **MOUNTAIN WHITETHORN**, *Ceanothus cordulatus*. **[conif. chap.]** 3-5' high shrub with very thick branch system; branchlets rigid and thorny; smooth, white bark; leaves evergreen and leathery, light to gray-green above, paler below; sweet-smelling white flowers. Common in the red fir forest.

40. **ROTHROCK SAGEBRUSH**, *Artemisia rothrockii*; Asteraceae, Sunflower fam. **[conif. chap. sub-alp.]** 8-19" high, erect shrub; dark green leaves and stems are sticky, sometimes whitened with a dense fuzz; the tiny flower heads (#11) have 10-16 yellowish or purplish disk flowers (no ray flowers). Foliage with pungent odor. Common in the lodgepole pine forest.

HERBS

41.☆ **WOOD STRAWBERRY**, *Fragaria vesca*; Rosaceae, Rose fam. **[conif. chap.]** Stems 2-11" long with thin, light green leaves and usually white flowers; berries bright red and edible.

42. **SCARLET GILIA**, *Ipomopsis aggregata*; Polemoniaceae, Phlox fam. **[conif. rocks, mead.]** Plant 4-20" with very thin, 9-11 lobed, pinnate leaves; flowers scarlet, tubular and funnel-form (#4) in a close panicle (#17); flowers may also be pink or yellow and often spotted with white or yellow. Tulare Co. north.

43

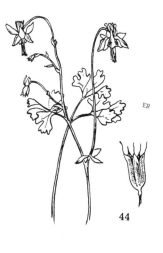

ER

44

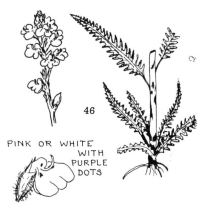

CY

45

ER

46

PINK OR WHITE
WITH
PURPLE
DOTS

CY

YELLOW

WHITE
WOOLY

ER

47

48

ER

43. **SNOW PLANT,** *Sarcodes sanguinea*; Ericaceae, Heath fam. **[conif.]** Entire plant red to orange-red, thick, fleshy, 1' tall. Feeds on humus.

44. **NORTHWEST CRIMSON COLUMBINE,** *Aquilegia formosa*; Ranunculaceae, Buttercup fam. **[conif. sub-alp. mead.]** Plant 10-30" tall with nodding, bright red flowers; basal leaves thin and in double sets of three leaflets; higher leaves much smaller and scattered. Generally in moist woods; several varieties.

45. **SHOWY PENSTEMON,** *Penstemon speciosus*; Scrophulariaceae, Figwort fam. **[conif. sage, mead.]** Plant 2-24" tall with thick, often spoon-shaped leaves; showy, bright purple-blue flowers sympetalous (#10) and irregular, in clusters. Usually east side of mountains.

46. **LITTLE ELEPHANT'S HEAD,** *Pedicularis attolens*. **[conif. mead.]** Plant 6-18" tall with basal, toothed, nearly linear, pinnately-divided leaves; flowers, light pink to lavender with purple dots, look like tiny elephant heads, each with a trunk. From San Bernardino Mts. north. **ELEPHANT'S HEAD,** *P. groenlandica* is similar to above, but its flowers are larger and red-purple.

47. **SIERRA MULE EARS,** *Wyethia mollis*; Asteraceae, Sunflower fam. **[conif. mead. rocks]** Plant 12-19" with large, basal, earlike leaves and smaller leaves on stem, all covered with woolly-white hair, often becoming hairless; ray flowers (#11) yellow and quite large. From Mariposa Co. north.

48. **LODGEPOLE PINE SENECIO,** *Senecio integerrimus* var. *major*. **[conif. mead.]** Plant 11-27" tall with a few-leaved stem rising from densely clustered roots; yellow flowers in heads (#11) surrounded by black tipped bracts, the heads forming cymes (#15); 10-12 conspicuous yellow ray flowers. Woods, moist ground from Tulare Co. north.

Other Plants Found in the Middle Mountain Forests

Sierra Water Fern,	11	Cinquefoils,	29, 37, 45	Pussy Paws,	37
Lady Fern,	11	Sulphur Flower,	31	Red Larkspur,	37
Five-finger Fern,	11	Western Bistort,	31	Harvest Brodiaea,	37
Leopard Lily	13	Sierra Stonecrop,	31	Fireweed,	39
Quaking Aspen,	13	Sedges,	35	Nettle-leaved Horsemint,	39
Willows,	13	Long-styled Rush,	35	Western Dog Violet,	39
Mountain Dogwood,	13	Red Fescue,	35	Torrey's Lupine,	39
Maples,	13, 14	One-sided Bluegrass,	35	Sierra Shooting Star,	39
Western Azalea,	14	Calif. Needlegrass,	35	Indian Paintbrush,	39
Box Elder,	14	Golden Brodiaea,	35	Primrose Monkeyflower,	41
Limber Pine,	27	Scabrous Bent Grass,	35	Bigelow Sneezeweed	41
Whitebark Pine,	27	Idaho Bent Grass,	35	Sierra Bristly Aster,	41
Foxtail Pine,	29	Corn Lily,	35	American Parsley Fern,	47
Wax Currant,	29	Mariposa Lily,	37	Cliff Brake,	47
Mountain Gooseberry,	29	Nude Buckwheat,	37		
Hoary Buckwheat,	29	Monkshood,	37		

Common Animals of the Middle Mountain Forests

Mammals

Little Brown Bat,	57	Cooper's Hawk,	81	Mountain Bluebird,	99
California Bat,	57	Sharp-shinned Hawk,	81	Townsend's Solitaire,	101
Long-eared Bat,	57	Red-tailed Hawk,	81	Golden-crowned Kinglet,	101
Hairy-winged Bat,	57	Golden Eagle,	81	Ruby-crowned Kinglet,	101
Big Brown Bat,	57	Sparrow Hawk,	83	Solitary Vireo,	101
Silver-haired Bat,	57	Blue Grouse,	83	Orange-crowned Warbler,	103
Red Bat,	57	Mountain Quail,	83	Nashville Warbler,	103
Hoary Bat,	57	Band-tailed Pigeon,	85	MacGillivray's Warbler,	103
Black Bear,	59	Great Horned Owl,	85	Black-throated Gray Warbler,	103
Raccoon,	59	Flammulated Owl,	85	Yellow-rumped Warbler,	103
Pine Marten,	59	Northern Pygmy Owl,	87	Hermit Warbler,	103
Fisher,	59	Spotted Owl,	87	Brewer's Blackbird,	105
Short-tailed Weasel,	59	Great Gray Owl,	87	Western Tanager,	105
Long-tailed Weasel,	59	Calliope Hummingbird,	89	Black-headed Grosbeak,	105
Mink,	59	Broad-tailed Hummingbird,	87	Chipping Sparrow,	107
Wolverine,	61	Common Nighthawk,	87	Oregon Junco,	109
Striped Skunk,	61	Northern Flicker,	89	Evening Grosbeak,	109
Spotted Skunk,	61	Pileolated Woodpecker,	89	Purple Finch,	109
Red Fox,	61	Acorn Woodpecker,	89	Cassin's Finch,	109
Coyote,	61	Lewis' Woodpecker,	91	Pine Grosbeak,	111
Bobcat,	61	Hairy Woodpecker,	91	Red Crossbill,	111
Mountain Lion,	61	White-headed Woodpecker,	91	Lesser Goldfinch,	111
Golden-mantled Squirrel,	63	Williamson's Sapsucker,	91	Pine Siskin,	111
Lodgepole Chipmunk,	63	Red-breasted Sapsucker,	91		
Sonoma Chipmunk,	63	Pacific Slope Flycatcher,	91	**Reptiles**	
Long-eared Chipmunk,	65	Gray Flycatcher,	93	Sagebrush Lizard,	115
Shadow Chipmunk,	65	Western Wood Pewee,	93	Western Skink,	115
Chickaree,	65	Olive-sided Flycatcher,	93	Gilbert's Skink,	115
Western Gray Squirrel,	65	Hammond's Flycatcher,	93	Southern Alligator Lizard,	115
Northern Flying Squirrel,	65	Tree Swallow,	93	Northern Alligator Lizard,	115
Deer Mouse,	67	Violet-green Swallow,	93	Rubber Boa,	115
Brush Mouse,	67	Purple Martin,	93	Calif. Mountain Kingsnake,	117
Piñon Mouse,	67	Steller's Jay,	95	W. Yellow-bellied Racer,	117
Western Harvest Mouse,	67	American Crow,	95	Pacific Ringneck Snake,	117
Dusky-footed Woodrat,	67	Clark's Nutcracker,	95		
Bushy-tailed Woodrat,	69	Chestnut-backed Chickadee,	97	**Amphibians**	
Mountain Meadow Vole,	69	Mountain Chickadee,	95	Rough-skinned Newt,	121
Calif. Red-backed Vole,	69	White-breasted Nuthatch,	97	Sierra Newt,	121
Porcupine,	71	Red-breasted Nuthatch,	97	S. Long-toed Salamander,	121
Snowshoe Rabbit,	71	Pygmy Nuthatch,	97	Pacific Giant Salamander,	123
Mule Deer,	72	Brown Creeper,	97	Sierra Nevada Salamander,	123
Elk,	72	House Wren,	97	Calif. Slender Salamander,	123
		Winter Wren,	99	Western Toad,	125
Birds		American Robin,	99	Pacific Treefrog,	125
Turkey Vulture,	81	Hermit Thrush,	99		
Northern Goshawk,	81	Varied Thrush,	99		
		Western Bluebird,	99		

THE SUB-ALPINE FOREST

Hudsonian and Arctic-alpine zones in mountains. Courtesy Southern Pacific .

This is the highest forest in the region, going to the edge of timberline where all trees are stunted and misshapen by the wind (about 8000–9500' in Northern California, and about 9500–11000' in the Sierra Nevada). It is a cold, harsh region, with very few frost-free days, and much snow. The meadows of this region are so intimately associated with the rather scattered trees of this forest that they are here considered part of it. Most of the grasses and the flowers described in the alpine meadow section (page 43) are also found here.

TREES

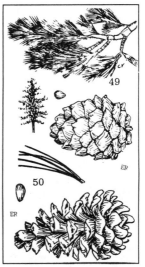

49. **WHITEBARK PINE**, *Pinus albicaulis*; Pinaceae, Pine fam. [sub-alp. conif.] 15-50' high evergreen; usually stunted and low with a crooked or even lying-down trunk; the branches short and thick; the thin bark is divided into gray-white scales; the stiff dark green needles are in 5's, 1½-3" long; purplish brown cones 1½-3½" long.

50. **LIMBER PINE**, *Pinus flexilis*. [sub-alp. conif.] Up to 65' high evergreen with a short, stout trunk; young bark thin and light gray; old bark breaks into dark brown plates; dark green needles in 5's, marked with little pits; cones yellow-brown, 3-10" long. In Warner Mts. and eastern slope Sierra from El Dorado Co. south.

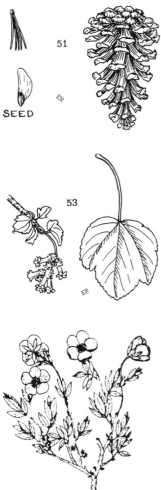

51

SEED

52

53

54

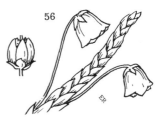

56

55

CY

57

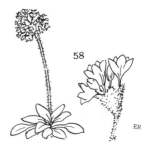

58

51. **FOXTAIL PINE,** *Pinus balfouriana.* **[sub-alp. conif.**] 20-50′ high evergreen with stout, short branches; the branchlets are crowded with thick clusters of the dark, blue-green needles (in 5′s), whitish on the insides; cones dark red-brown, 2-7″ long, covered with incurved prickles. Inner northern Coast Ranges and southern Sierra Nevada.

SHRUBS

52. **ARCTIC WILLOW,** *Salix arctica*; Salicaceae, Willow fam. **[sub-alp. alpine]** The stems creep over the ground, usually rising only about 4″; brownish in color; 3/4-1½″ long leaves deep green above, paler beneath. Found from Inyo and Tulare counties north in alpine tundra..

53.☆ **WAX CURRANT,** *Ribes cereum*; Grossulariaceae, Gooseberry fam. **[sub-alp. alpine, pin-jun. conif.**] 1½-3½′ high shrub with many branches, the branchlets covered with tiny hairs; white to pink tubular flowers in short, hanging racemes (#16); berries bright red, edible. Indians made a jelly from the berries. Dry rocky places.

54.☆ **MOUNTAIN GOOSEBERRY,** *Ribes montigenum.* **[sub-alp. alpine, conif.**] 1-2′ high spreading bush with the stems usually bristly; leaves and most of the plant covered with dense, short hairs often sticky to touch; racemes (#16) 3-7 flowered; flowers red to purplish; berries orange-red, covered with glandular bristles; edible. Dry rocky places.

55. **SHRUBBY CINQUEFOIL,** *Potentilla fruticosa*; Rosaceae, Rose fam. **[sub-alp. alpine, mead. conif.**] ½-3′ high undershrub with many branches and very leafy stems; the brown bark shreds off; leaflets entire, hairy-silky; flowers yellow. Moist places.

56. **WHITE HEATHER,** *Cassiope mertensiana*; Ericaceae, Heath fam. **[sub-alp. alpine, rocks]** 4-12″ high; a very low, densely branched bush; tiny evergreen leaves leathery, appressed to stems, keeled; solitary bell-shaped flowers white to pinkish; many tiny, winged seeds. From Fresno and Trinity counties north.

57. **PURPLE MOUNTAIN HEATHER,** *Phyllodoce breweri.* **[sub-alp. alpine, rocks]** 4-16″ high, rigid and erect shrub with linear evergreen leaves (#31); the campanulate (#3) flowers are deep rose purple or pinkish.

HERBS

58. **HOARY BUCKWHEAT,** *Eriogonum marifolium*; Polygonaceae, Buckwheat fam. **[sub-alp. rocks, conif.**] Plant 4-15″ tall covered with dense white or hoary hairs; the pale, lemon yellow flowers appear in a tiny umbel(#12), almost like a head (#11),

59

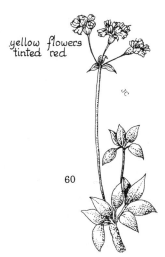

yellow flowers
tinted red

60

61

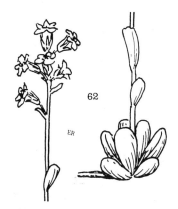

62

63

surrounded by leaflike bracts. Found mainly from Fresno Co. north on dry gravelly slopes.

59. **WESTERN BISTORT,** *Polygonum bistortoides*. **[sub-alp. strwd. conif. mead.]** Stems 8-25" long; several erect, hairless, unbranched stems rise from a thick, level rootstock; spikes pink or white. Wet meadows and along streams.

60. **SULPHUR FLOWER,** *Eriogonum umbellatum* var. *polyanthym*. **[sub-alp. sage, conif.]** Flowering stems 4-12" high, few to many branches from woody base; leaves white hairy beneath; flowers sulphur-yellow becoming reddish, in ball-like clusters. Dry slopes.

61. **COVILLE'S COLUMBINE,** *Aquilegia pubescens*; Ranunculaceae, Buttercup fam. **[sub-alp. alpine, rocks]** Plant 5-19" with solitary, showy, cream to pink flower, standing erect instead of nodding (as in most columbines); basal stem leaves generally deeply 3-lobed, often covered with tiny hairs. Rocky places. Found from Tuolumne Co. to Tulare and Inyo counties.

62. **SIERRA STONECROP,** *Sedum obtusatum*; Crassulaceae, Stonecrop fam. **[sub-alp. alpine, conif. rocks]** Stems 1-8" long rise from a thick, horizontal rootstock; basal leaves (#27) spoon-like, thick and fleshy and pale green; petals yellow or cream-colored, united only near their bases. Rocky slopes mostly in the lodgepole forest. Tulare to Plumas counties; also Trinity Co. north.

63. **SIERRA PENSTEMON,** *Penstemon heterodoxus*; Scrophulariaceae, Figwort fam. **[sub-alp. alpine, rocks, mead.]** Plant 3-25" with slender stems; leaves thin, dark green; flowers deep blue-purple, tipped with brownish yellow beards. From Tulare Co. to Plumas Co.

Other Plants of the Sub-alpine Forest and Meadows

Lady Fern,	11	NW Crimson Columbine,	25	Bigelow Sneezeweed,	41
Alpine Lily,	11	Sedges,	35	Sierra Bristly Aster,	41
Common Brake Fern,	17	Long-styled Rush,	35	Douglas' Phlox,	47
Jeffrey Pine,	17	Corn Lily,	35		
Western White Pine,	19	Hairy Cinquefoil,	37		
Lodgepole Pine,	19	Pussy Paws,	37		
Mountain Hemlock,	19	Indian Paintbrush,	39		
Mountain Pink Currant,	21	Sierra Shooting Star,	39		
Rothrock Sagebrush,	23	Primrose Monkeyflower,	41		

Common Animals of the Sub-alpine Forest and Meadows

Mammals

Vagrant Shrew,	55
Mount Lyell Shrew,	55
Ornate Shrew,	55
Shrew-mole,	55
Little Brown Bat,	57
Black Bear,	59
Pine Marten,	59
Short-tailed Weasel,	59
Long-tailed Weasel,	59
Badger,	59
Wolverine,	61
Red Fox,	61
Coyote,	61
Mountain Lion,	61
Yellow-bellied Marmot,	63
Golden-mantled Squirrel,	63
Belding's Ground Squirrel,	63
Lodgepole Chipmunk,	63
Alpine Chipmunk,	63
Chickaree,	65
Mountain Pocket Gopher,	65
Deer Mouse,	67
Bushy-tailed Woodrat,	69
Mountain Meadow Vole,	69
Long-tailed Vole,	69
Mountain Phenacomys,	69
Pacific Jumping Mouse,	69
Mountain Beaver,	71
Porcupine,	71
Pika,	71
White-tailed Hare,	71
Snowshoe Rabbit,	71
Mule Deer,	72
Bighorn Sheep,	72

Birds

Turkey Vulture	81
Northern Goshawk,	81
Cooper's Hawk,	81
Red-tailed Hawk,	81
Golden Eagle,	81
Sparrow Hawk,	83
Blue Grouse,	83
Great Horned Owl,	85
Northern Saw-whet Owl,	87
Great Gray Owl,	87
Common Nighthawk,	87
Calliope Hummingbird,	89
Northern Flicker,	89
Pileolated Woodpecker,	89
Hairy Woodpecker,	91
White-headed Woodpecker,	91
Black-backed Woodpecker,	91
Williamson's Sapsucker,	91
Red-breasted Sapsucker,	91
Hammond's Flycatcher,	93
Gray Flycatcher,	93
Western Wood Pewee,	93
Olive-sided Flycatcher,	93
Horned Lark,	95
Steller's Jay,	95
Common Raven,	95
Clark's Nutcracker,	95
Mountain Chickadee,	95
White-breasted Nuthatch,	97
Red-breasted Nuthatch,	97
Brown Creeper,	97
House Wren,	97
Winter Wren,	99
American Robin,	99

Hermit Thrush,	99
Western Bluebird,	99
Mountain Bluebird,	99
Townsend's Solitaire,	101
Golden-crowned Kinglet,	101
Ruby-crowned Kinglet,	101
Water Pipit,	101
Warbling Vireo,	101
Orange-crowned Warbler,	103
Yellow-rumped Warbler	103
Hermit Warbler,	103
MacGillivray's Warbler,	103
Wilson's Warbler,	105
Western Tanager,	105
Chipping Sparrow,	107
Oregon Junco,	109
Evening Grosbeak,	109
Cassin's Finch,	109
Red Crossbill,	111
Pine Grosbeak,	111
Pine Siskin,	111

Reptiles

Sagebrush Lizard,	115
N. Pacific Rattlesnake,	119

Amphibians

Mt. Lyell Salamander,	123
Pacific Treefrog,	125

MIDDLE MOUNTAIN MEADOWS

Photo courtesy of U.S. Forest Service.

These are the meadows associated with the middle mountain forests already described. They lie deep under snow in the winter, but, with the coming of the warm spring and summer days, they often turn into veritable paradises of wildflowers, attracting birds and other animals to their lush greenness. When overgrazed, however, by cattle or sheep, they become very poor-looking, and serve as a lesson for the need to intelligently conserve our natural resources.

GRASSES

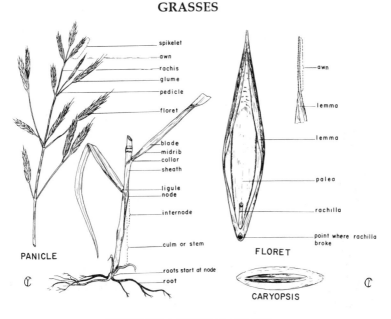

PANICLE

spikelet
awn
rachis
glume
pedicle
floret
blade
midrib
collar
sheath
ligule
node
internode
culm or stem
roots start at node
root

FLORET

awn
lemma
lemma
palea
rachilla
point where rachilla broke

CARYOPSIS

PARTS OF A TYPICAL GRASS
To help identify grasses the picture to the left shows parts of a complete grass. The picture to the right shows parts of a floret.

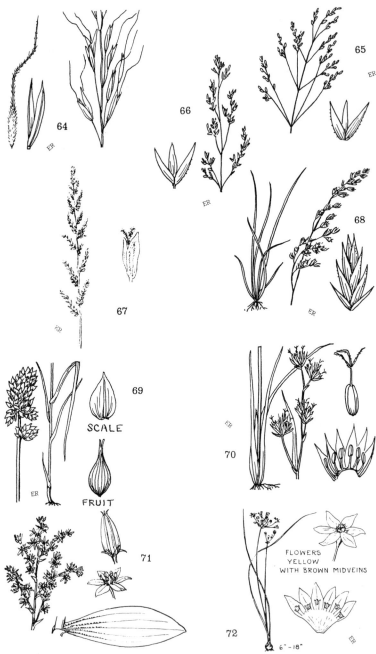

64. **CALIFORNIA NEEDLEGRASS,** *Achnatherum occidentalis* ssp. *californicum*; Poaceae, Grass fam. [mead. conif.] 1-4' high grass with floret awns bent in two places; the sheaths that protect the florets are smooth and the leaf blades are flat becoming rolled.

65. **SCABROUS BENT GRASS,** *Agrostis scabra.* [mead. conif.] 8-32" tall grass with delicate slender stems and a very diffuse panicle (#17); lower leaf blades flat and scabrous (rough feeling).

66. **IDAHO BENT GRASS,** *Agrostis idahoensis.* [mead. conif.] 4-12" high grass with slender stems; a loosely spreading panicle, though not as much as in the above grass, and shorter leaf blades.

67. **ONE-SIDED BLUEGRASS,** *Poa secunda* ssp. *secunda.* [mead. conif. str-wd.] 6-39" tall grass with rough-feeling stems and floret sheaths; leaf blades mainly basal (#27); panicle linear to lanceolate, generally dense and more or less 1-sided, with branches appressed to ascending (spreading only in flowering). There are many kinds of bluegrasses in the Sierras. To 5000'.

68. **RED FESCUE,** *Festuca rubra.* [mead. conif. str-wd.] 1-2½' high grass with the lower leaf sheaths often reddish and always smooth, shedding with age; the leaf blades feel soft and are usually rolled inwards; the pale green spikelets often purplish tinged. Up to 8500'.

69. **SEDGES,** *Carex* species; Cyperaceae, Sedge fam. [mead. conif. sub-alp. str-wd. water] Very numerous species, all characterized by triangular-shaped stems, and long, narrow leaves in 3 ranks; the stems form one to many spikes. Very common in damp meadows at all altitudes.

70. **LONG-STYLED RUSH,** *Juncus longistylis*; Juncaceae, Rush fam. [mead. conif. sub-alp.] 8-23" high with erect, loosely tufted stems, circular in cross-section; the flat, grasslike basal leaves have rough sheaths over the stem; florets greenish brown with transparent margins. There are many kinds of rushes in the mountains, all similar to this rush.

HERBS

71. **CORN LILY** or **FALSE HELLEBORE,** *Veratrum californicum*; Liliaceae, Lily fam. [mead. sub-alp. str-wd. conif.] Stem 3-6' long, hollow; leaves large, ovate, parallel-veined (#35); flowers white with greenish veins, in panicles (#17). Up to 10,000'.

72.☆ **GOLDEN BRODIAEA,** *Triteleia ixioides* ssp. *scabra.* [mead. conif. chap.] Stem 8-30" high with 2-3 linear (#31), basal (#27) leaves; flowers yellow marked with brown midveins in an open umbel (#12). Anthers of stamens oval and creamy white to blue. From Butte Co. south.

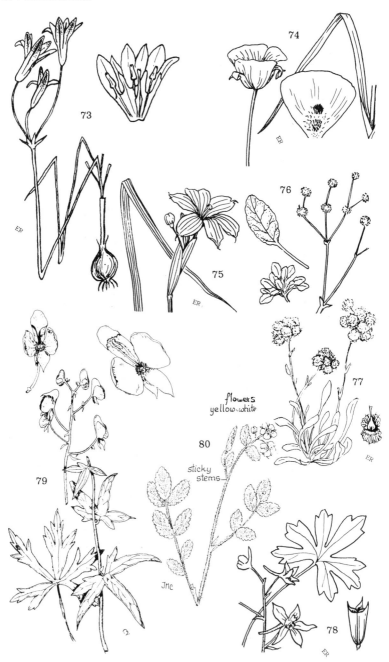

73

74

75

76

77

78

79

80

flowers
yellow-white

sticky
stems

73.☆ **HARVEST BRODIAEA,** *Brodiaea elegans.* **[mead. conif.]** Stem 4-15" tall with narrow leaves about as long as stem; flowers blue-purple to violet and funnel-form (#4); umbels (#12) with 2-11 flowers. Like other brodiaeas it rises from an edible white bulb sheathed in a brown, fibrous cover. Mainly found in the lower meadows of ponderosa pine forest.

74. **MARIPOSA LILY,** *Calochortus venustus.* **[mead. conif.]** Stem 4-24" high, erect and generally branched; umbels of 1-3 white, yellow, purple or dark red flowers, each petal with a dark eye-spot in the middle and often a second paler blotch above the first; 1-2 basal leaves. Lower meadows from Tulare Co. north.

75. **BLUE-EYED GRASS,** *Sisyrinchium bellum;* Iridaceae, Iris fam. **[mead. str-wd.]** Stem 4-20" high with grasslike leaves as long as stem from base, plus smaller leaves higher up; flowers deep bluish purple to blue-violet, or pale blue, each petal tipped with a point. In lower meadows.

76. **NUDE BUCKWHEAT,** *Eriogonum nudum;* Polygonaceae, Buckwheat fam. **[mead. conif. rocks, chap.]** 8-40" high, but the high Sierra form is shorter; a perennial with spreading, basal (#27) leaves covered with white fuzz below; white flowers in heads with the sepals (#2) having rose-colored veins. Usually on dry or rocky slopes.

77. **PUSSY-PAWS,** *Calyptridium umbellatum;* Portulacaceae, Purslane fam. **[mead. conif. sub-alp.]** Less than 2' high herb with spreading stems, often hugging the ground; leaves spoon-shaped, basal, sometimes small leaves on stems; the fuzzy pink to white flowers form dense clusters at ends of reddish stalks. Sandy soil.

78. **RED LARKSPUR,** *Delphinium nudicaule;* Ranunculaceɑ Buttercup fam. **[mead. conif. chap.]** Stem generally 6-20" long, hairless leaves mostly on lower 30% of stem; flowers in a loose raceme (#16), upper petals yellow, tipped with red, sepals scarlet to orange-red, sometimes yellow; seeds winged. Found mainly from Mariposa Co. north.

79. **MONKSHOOD,** *Aconitum columbianum.* **[mead. conif.]** Stem 1-5' tall; all leaves palmately lobed or divided, 3-5 times; flowers purplish blue, sometimes white, in a loosely few-flowered raceme; the uppermost sepal forms a hood, enclosing the upper 2 petals; fruit of 2-6 follicles or pods. Found in moist areas, especially near willows.

80. **HAIRY CINQUEFOIL,** *Potentilla glandulosa;* Rosaceae, Rose fam. **[mead. conif. sub-alp. chap.]** 4-35" high with several very hairy stems which rise from a short, woody trunk; the

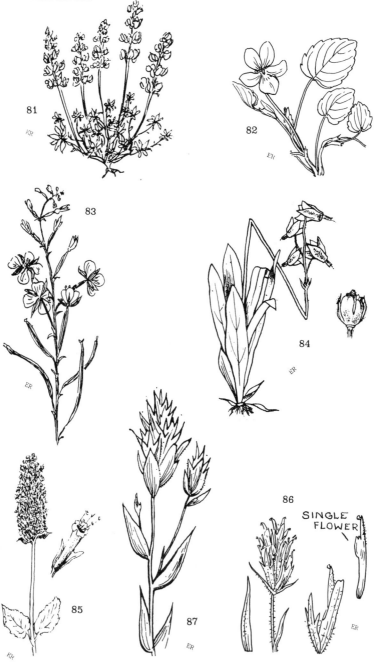

81

82

83

84

85

86

87

SINGLE
FLOWER

basal leaves (#27) are short-stemmed and covered with very dense, short, silky hair; 5-9 leaflets are in each leaf; yellow to white flowers with 5 petals. Found mainly at low elevations.

81. **TORREY'S LUPINE**, *Lupinus lepidus* var. *sellulus*; Fabaceae, Pea fam. [**mead. conif. sage**] 5-10" high herb with crowded basal leaves and no woody base; 6-8 leaflets with appressed, silky hairs; flowers violet to blue with the center of the top petal (or banner) touched with white to yellow. From Mariposa Co. north.

82. **WESTERN DOG VIOLET**, *Viola adunca*; Violaceae, Violet fam. [**mead. conif. str-wd.**] 2-12" herb rising from a narrow, branching rootstock; basal leaves with low, broad teeth; petals pale to deep violet, lower three white at base and purple veined, lateral two white-bearded, often hooked at tip.

83. **FIREWEED**, *Epilobium angustifolium*; Onagraceae, Evening Primrose fam. [**mead. chap. conif.**] 2-8' high perennial herb with 1 or few slender, erect stems; leaves lance-shaped with the side-veins forming loops along the edges of the leaves; flowers deep rose-pink clustered in racemes. Found in open places, especially after fires.

84. **SIERRA SHOOTING STAR**, *Dodecatheon jeffreyi*; Primulaceae, Primrose fam. [**mead. sub-alp. alpine, conif.**] Leaves 4-19" long; distinctive reversed petals magenta to lavender or white with maroon anthers (#1d), dark band at throat. **ALPINE SHOOTING STAR**, *D. alpinum*. Similar, but has smaller, shorter leaves.

85. **NETTLE-LEAVED HORSEMINT**, *Agastache urticifolia*; Lamiaceae, Mint fam. [**mead. conif. str-wd.**] Generally less than 3' high herb with several erect stems ending in distinctive spikes (#14) of rose to rose-purple flowers; coarsely toothed leaves 1-3½", shiny, sometimes irritating to skin.

86. **INDIAN PAINTBRUSH**, *Castilleja lemmonii*; Scrophulariaceae, Figwort fam. [**mead. sub-alp. alpine, str-wd.**] 4-8" high perennial covered with tiny hairs; somewhat sticky or glandular on lower parts; numerous stems; the bracts surrounding the flowers are purplish red, the flowers green with thin purple margins. Moist places, generally above 7,000 feet.

87. **GREAT RED INDIAN PAINTBRUSH**, *Castilleja miniata*. [**mead. conif.**] 16-32" high perennial, smooth or hairy; the stems form clusters; the bracts surrounding the flowers are usually bright red, but sometimes yellowish; petals green with thin red margins, usually ¾-1¼" long; flowers (as in most paintbrushes)

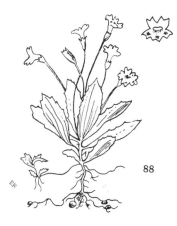

88

89

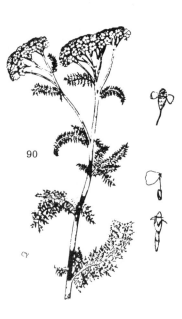

90

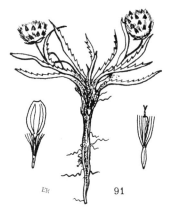

91

92

in spikelike racemes. There are many varieties of paintbrushes in our mountains.

88. **PRIMROSE MONKEYFLOWER,** *Mimulus primuloides.* **[mead. str-wd. conif.]** This is just one among many kinds of monkeyflowers in the mountains. Up to 5" high, usually with a few flower stems rising from a basal rosette of leaves; flowers ½-¾" long, yellow, funnel-form, the palate deeper yellow and very hairy, spotted red-brown, or with one spot below. Forms mats in damp meadows.

89. **BIGELOW SNEEZEWEED,** *Helenium bigelovii*; Asteraceae, Sunflower fam. **[mead. conif. sub-alp.]** Stems 1-4' high, generally branching; leaves up to 8" long, gland dotted sometimes with very tiny hairs; the yellow ray flowers (#11) droop down from the yellow to red, brown or purple ball of disk flowers. May cause sneezing when pollen is breathed.

90. **COMMON YARROW,** *Achillea millefolium.* **[mead.]** ½-6' high perennial with a simple stem and leaves 3-pinnately divided into numerous fine segments; ¼-½" wide heads in corymbs (#13) at end of stem; each head with a few white to pink ray flowers (#11) surrounding white to purple or yellow disk flowers.

91. **SIERRA BRISTLY ASTER,** *Pyrrocoma apargioides* **[mead. rocks, conif. sub-alp. alpine]** 3-11" high herb with several stems rising from a very thick root, which is sheathed in old leaves; leaves mainly in a basal tuft, 1½-4" long; heads (#11) 1-3, usually surrounded by purplish or green-tipped bracts; both ray and disk flowers yellow; seeds surrounded by dirty yellow bristles.

92. **CALIFORNIA CONE FLOWER,** *Rudbeckia californica.* **[mead.]** 2-6' tall herb with generally unbranched stem topped by a solitary head with showy, yellow ray flowers and greenish yellow disk flowers. **BLACK-EYED SUSAN,** *R. hirta.* Similar, except with dark purple disk flowers and rough, hairy, and generally branching stems.

Common Animals of the Middle Mountain Meadows

Mammals

Broad-footed Mole,	55
Shrew-mole,	55
Vagrant Shrew,	55
Ornate Shrew,	55
Trowbridge Shrew,	55
Montane Shrew,	55
Little Brown Bat,	57
California Bat,	57
Long-eared Bat,	57
Small-footed Bat,	57
Big Brown Bat,	57
Red Bat,	57
Hoary Bat,	57
Silver-haired Bat,	57
Black Bear,	59
Pine Marten,	59
Short-tailed Weasel,	59
Long-tailed Weasel,	59
Mink,	59
Badger,	59
Wolverine,	61
Striped Skunk,	61
Spotted Skunk,	61
Red Fox,	61
Coyote,	61
Bobcat,	61
Mountain Lion,	61
Yellow-bellied Marmot,	63
California Ground Squirrel,	63
Golden-mantled Squirrel,	63
Belding's Ground Squirrel,	63
Mountain Pocket Gopher,	65
Botta's Pocket Gopher,	67
Deer Mouse,	67
Western Harvest Mouse,	67
Bushy-tailed Woodrat,	69
Mountain Meadow Vole,	69
Oregon Vole,	69
Long-tailed Vole,	69
Pacific Jumping Mouse,	69
Pika,	71
Black-tailed Hare,	71
White-tailed Hare,	71
Mule Deer,	72
Elk,	72

Birds

Great Blue Heron,	77
Canada Goose,	77
Northern Goshawk,	81
Sharp-shinned Hawk,	81
Golden Eagle,	81
Swainson's Hawk,	83
Northern Harrier,	83
Sparrow Hawk,	83
Mountain Quail,	83
Mourning Dove,	83
Killdeer,	85
Great Horned Owl,	85
Common Nighthawk,	87
Rufous Hummingbird,	87
Allen's Hummingbird,	87
Broad-tailed Hummingbird,	87
Calliope Hummingbird,	89
Northern Flicker,	89
Black Swift,	89
Tree Swallow,	93
Purple Martin,	93
Violet-green Swallow,	93
American Crow,	95
Black-billed Magpie,	95
Bushtit,	97
American Robin,	99
Western Bluebird,	99
Mountain Bluebird,	99
Water Pipit,	101
Brewer's Blackbird,	105
Bullock's Northern Oriole,	105
Black-headed Grosbeak,	105
Vesper Sparrow,	107
Chipping Sparrow,	107
White-crowned Sparrow,	107
Lincoln's Sparrow,	109
Oregon Junco,	109
Pine Siskin,	111
Lesser Goldfinch,	111

Reptiles

Western Fence Lizard,	115
Sagebrush Lizard,	115
Western Skink,	115
Gilbert's Skink,	115
Calif. Mountain Kingsnake,	117
Common Kingsnake,	117
W. Yellow-bellied Racer,	117
Pacific Gopher Snake,	119
N. Pacific Rattlesnake,	119

Amphibians

S. Long-toed Salamander,	121
Great Basin Spadefoot,	125
Western Toad,	125
Yosemite Toad,	125
Pacific Treefrog,	125
Foothill Yellow-legged Frog,	125
Red-legged Frog,	125
N. Leopard Frog,	126

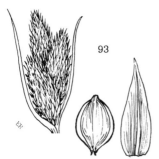

93

94

ALPINE FELL FIELDS OR MEADOWS

Photo courtesy of Sierra Club.

This is a land of ice, snow, and rocks above tree growth, a land where there is a brief and almost violent growth of wildflowers in mid-summer, and where only a few animals can find food to support themselves. The growing season lasts only a month to seven weeks. The sun is very bright in summer on the snow, but the temperature rarely goes above 60 degrees. Most of the common alpine plants described here are also found, though usually less commonly, in the sub-alpine meadows.

93. **HELLER'S SEDGE**, *Carex helleri*; Cyperaceae, Sedge fam. Sedges generally have solid, sharply 3-angled stems. Stem 5-19" high, in dense tufts. From Tulare Co. north.

94. **BREWER'S SEDGE**, *Carex breweri*. Stem 4-10" high; leaf blades rolled quill-like; a solitary spike (#14). From Mt. Whitney north.

95. **SMALL SHEEP FESCUE**, *Festuca brachyphylla*; Poaceae, Grass fam. Stems 2-8" high, densely clumped; flower spikes rough to touch.

96. **TIMBERLINE BLUEGRASS**, *Poa glauca* ssp. *rupicola*. Stems 2-6" high, stiff in dense tufts; spikelets in compact panicle, usually purplish. Central and south high Sierras.

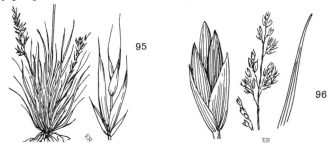

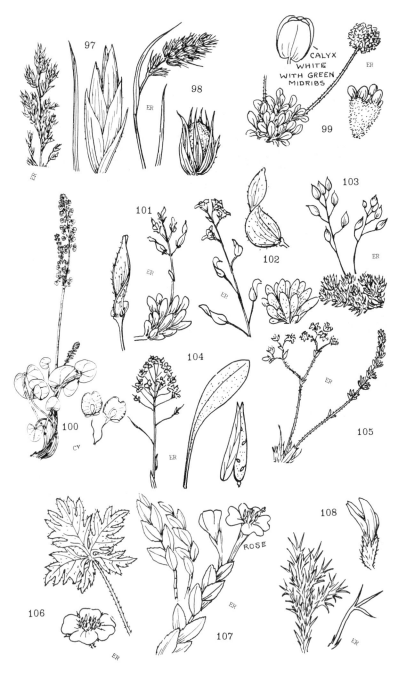

97. **KECK'S BLUEGRASS**, *Poa keckii*. Generally 1½-4" high; densely clumped; leaf-sheaths loose, paperlike, and smooth. From Ventura Co. north.

98. **SPIKED WOOD RUSH**, *Luzula spicata*; Juncaceae, Rush fam. 1-12" high; densely clumped; leaves are stiffly erect; panicle of dense, brown, nodding spike-like flower clusters; linear channels appear in leaves. From Tulare Co. north.

99. **OVAL-LEAVED BUCKWHEAT**, *Eriogonum ovalifolium* var. *nivale*; Polygonaceae, Buckwheat fam. Flowering stem less than 2½" high; white woolly all over; very numerous basal leaves; petals white with red veins. From Tulare Co. north.

100. **MOUNTAIN SORREL**, *Oxyria digyna*. Stem less than 19" high; the greenish flowers often have red sepals (#2); in compact panicles; juice acidic; flat fruit conspicuously winged.

101. **BREWER'S DRABA**, *Draba breweri*; Brassicaceae, Mustard fam. Stems less than 6", usually several rising from cushiony base; grayish leaves basal, densely hairy; petals white. From Tulare Co. north.

102. **LEMMON'S DRABA**, *Draba lemmonii*. Stems up to 6"; often forms mats; very compact leaf bunches; petals yellow; usually in rocks.

103. **WHITE-MOUNTAIN DRABA**, *Draba oligosperma*. Stems up to 5" form cushiony mats; leaves leatherlike and densely overlapping on stem; petals yellow. Found in central Sierra Nevada and White Mts.

104. **BROAD-PODDED ROCK FLOWER**, *Anelsonia eurycarpa*. Stems 1-2" high, covered all over with short hairs; basal leaves ⅖-⅔" long, oblanceolate, in dense clusters; flowers with small white or yellow petals; silvery seeds.

105. **SHOCKLEY'S IVESIA**, *Ivesia shockleyi*; Rosaceae, Rose fam. Stem spreading, up to 6" high; pallid green, delicate-looking plant, densely covered with tiny glandular hairs; 5-10 pairs leaflets; usually in 10-flower arrangements; petals yellow; 2-5 pistils. North and central Sierra Nevada.

106. **DIVERSE-LEAVED CINQUEFOIL**, *Potentilla diversifolia*. Slender ascending stems 4-15"; leaves sparsely hairy, palmate; petals yellow. Central and south Sierra Nevada.

107. **ROCK FRINGE**, *Epilobium obcordatum*; Onagraceae, Evening Primrose fam. Plant 2-6", densely clumped; flowers pink to rose-purple. From Tulare Co. north.

108. **SPINY-LEAFED MILKVETCH**, *Astragalus kentrophyta* var. *tegetarius*; Fabaceae, Legume fam. More or less a flat mat; leaflets rigid and spine-tipped; petals usually purple, sometimes whitish. From Tulare Co. north.

109. **SIERRA PODISTERA**, *Podistera nevadensis*; Apiaceae, Carrot fam. Plant forms compact cushions 1-2" high; many stems from woody trunk; flower yellow. North and central Sierra.

110. **SKY PILOT**, *Polemonium eximium*; Polemoniaceae, Phlox fam. Stem up to 15"; leaves and stems glandular-sticky with musky odor; blue funnel-like flowers. From Tulare to Tuolumne Co.

111. **SIERRA CRYPTANTHA**, *Cryptantha nubigena*; Boraginaceae, Borage fam. Plant 1-11" high arising from a dense, leafy and branched root-crown; hairy and bristly all over; white flowers. Central and south Sierra Nevada (east slope particularly).

112. **DWARF ALPINE INDIAN PAINTBRUSH**, *Castilleja nana*; Scrophulariaceae, Figwort fam. Plant 2-10" high; irregular flowers pale yellow, blotched purplish; bracts yellow-green or purplish.

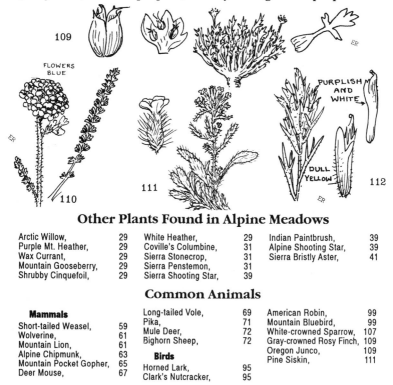

Other Plants Found in Alpine Meadows

Arctic Willow,	29	White Heather,	29	Indian Paintbrush,	39
Purple Mt. Heather,	29	Coville's Columbine,	31	Alpine Shooting Star,	39
Wax Currant,	29	Sierra Stonecrop,	31	Sierra Bristly Aster,	41
Mountain Gooseberry,	29	Sierra Penstemon,	31		
Shrubby Cinquefoil,	29	Sierra Shooting Star,	39		

Common Animals

Mammals		Long-tailed Vole,	69	American Robin,	99
		Pika,	71	Mountain Bluebird,	99
Short-tailed Weasel,	59	Mule Deer,	72	White-crowned Sparrow,	107
Wolverine,	61	Bighorn Sheep,	72	Gray-crowned Rosy Finch,	109
Mountain Lion,	61			Oregon Junco,	109
Alpine Chipmunk,	63	**Birds**		Pine Siskin,	111
Mountain Pocket Gopher,	65	Horned Lark,	95		
Deer Mouse,	67	Clark's Nutcracker,	95		

ROCKY AREAS AND CLIFFS
of middle altitudes

113. **AMERICAN PARSLEY FERN,** *Cryptogramma acrostichoides*; Pteridaceae, Brake fam. **[rocks, conif.]** Fronds 4-12" long, both short-stalked and long-stalked, forming thick clusters. Scales rusty to dark brown, or striped; fertile leaves with yellow stems.

114. **CLIFF-BRAKE,** or **INDIAN'S DREAM,** *Aspidotis densum*. **[rocks, conif.]** 6000-8000' in Sierra Nevada, 1000-6500' in Coast Ranges. Many crowded fronds, 6-12" high; rhizome with shiny brown scales.

115. **DOUGLAS' PHLOX,** *Phlox douglassii* ssp. *rigida*; Polemoniaceae, Phlox fam. **[rocks, mead. sub-alp.]** Densely clumped perennial with woody base; stems 1½-3" high; leaves rigid, linear, sharp-tipped; flowers pink or pale lavender to white. Found mostly on east side of mts.

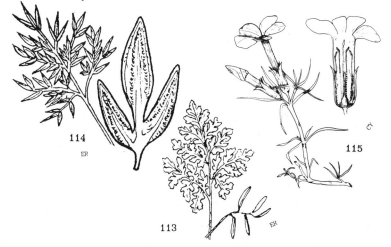

Other Plants of Rocky Areas

Five-finger Fern,	11	White Heather,	29	Nude Buckwheat,	37
Western Sword Fern,	17	Purple Mt. Heather,	29	Torrey's Lupine,	39
Common Brake Fern,	17	Hoary Buckwheat,	29	Sierra Bristly Aster,	41
Bush Chinquapin,	21	Sierra Stonecrop,	31	Most of Alpine plants,	43, 46
Scarlet Gilia,	23	Coville's Columbine,	31		
Sierra Mule Ears,	25	Sierra Penstemon,	31		

Common Animals of Rocks and Cliffs

Mammals

All Myotis Bats,	57
Big Brown Bat,	57
Ring-tailed Cat,	59
Pine Marten,	59
Short-tailed Weasel,	59
Long-tailed Weasel,	59
Wolverine,	61
Coyote,	61
Mountain Lion,	61
Bobcat,	61
Yellow-bellied Marmot,	63
Calif. Ground Squirrel,	63
Golden-mantled Squirrel,	63
Alpine Chipmunk,	63
Deer Mouse,	67

Brush Mouse,	67
W. Harvest Mouse,	67
Bushy-tailed Woodrat,	69
Pika,	71
Bighorn Sheep,	72

Birds

Turkey Vulture,	81
Golden Eagle,	81
Red-tailed Hawk,	81
Sparrow Hawk,	83
Great Horned Owl,	85
Black Swift,	89
White-throated Swift,	89
Belted Kingfisher,	89
Violet-green Swallow,	93
Common Raven,	95

Canyon Wren,	99
Rock Wren,	99
Gray-crowned Rosy Finch,	109

Reptiles

W. Fence Lizard,	115
Sagebrush Lizard,	115
N. Pacific Rattlesnake,	119

Amphibians

Mt. Lyell Salamander,	123
Limestone Salamander,	123
Black Salamander,	123
Shasta Salamander,	123

Piñon-Juniper Woodland

The piñon-juniper woodland habitat follows the east base of the Sierra Nevada from the White-Inyo ranges southward mostly at elevations of 5000 to 8000 feet, and also occurs between the ponderosa pine forest and joshua tree woodland or sagebrush scrub.

Plants

Western Juniper,	21
Wax Currant,	29

Mammals

Coyote,	61
Deer Mouse,	67
Piñon Mouse,	67
White-tailed Hare,	71
Bighorn Sheep,	72

Birds

Broad-tailed Hummingbird,	87
Calliope Hummingbird,	89
Gray Flycatcher,	93
Piñon Jay,	95
Black-billed Magpie,	95
Plain Titmouse,	95
Black-throated Gray Warbler,	103
California Towhee,	107

Reptiles and Amphibians

Western Skink,	115
Gilbert's Skink,	115
Desert Striped Whipsnake,	117
Great Basin Spadefoot,	125

In and Near Buildings

Deer Mouse,	67	Norway Rat,	71	House Sparrow,	111
Brush Mouse,	67	House Wren,	97	Western Fence Lizard,	115
House Mouse,	71				

CHAPARRAL AND SAGEBRUSH AREAS
In the Mountains

Chaparral in mountains; Photo courtesy U.S. Forest Service.

The chaparral or brushland habitat typically occurs on dry slopes and ridges in the Coast Ranges from Shasta County south and below the ponderosa pine forest on the western side of the Sierra Nevada. The sagebrush habitat typically runs along the east base of the Sierra Nevada from about 4000 to 7500 feet. But both these habitat types are found in pockets in the higher mountains, especially in dry rocky areas or areas recently burned over.

116. **BASIN SAGEBRUSH**, *Artemisia tridentata*; Asteraceae, Sunflower fam. **[sage, conif.]** In certain areas that are very dry this strong smelling, silvery gray bush occupies many acres. 2-10' high; 3-6 or even 12 flowers in a head; several heads on each panicle-branch. Sagebrush is often almost the only plant in the sagebrush habitat.

Burned over areas in the mountains temporarily spring up with brush, including such species as:

Common Plants

Common Brake Fern,	17	Hairy Cinquefoil,	37	
Bush Chinquapin,	21	Nude Buckwheat,	37	
Manzanita,	23	Red Larkspur,	37	
Deer Brush,	23	Fireweed,	39	
Mt. Whitethorn,	23	Torrey's Lupine,	39	
Rothrock Sagebrush,	23			
Wood Strawberry,	23			
*Showy Penstemon,	25			
*Sierra Mule Ears,	25			
Sulphur Flower,	31			
*Calif. Needlegrass,	35			
Golden Brodiaea,	35			

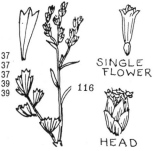

SINGLE FLOWER

116

HEAD

Animals and plants found most commonly in the sagebrush habitat are marked with an * in the lists above and below.

Common Animals of Chaparral and Sagebrush

Mammals

Ornate Shrew,	55
All Myotis Bats,	57
Big Brown Bat,	57
Black Bear,	59
Ring-tailed Cat,	59
Spotted Skunk,	61
Gray Fox,	61
Coyote,	61
Bobcat,	61
Mountain Lion,	61
Calif. Ground Squirrel,	63
Sonoma Chipmunk,	63
*Least Chipmunk,	65
Long-eared Chipmunk,	65
Merriam's Chipmunk,	65
Deer Mouse,	67
Brush Mouse,	67
West. Harvest Mouse,	67
Dusky-footed Woodrat,	67
Black-tailed Hare,	71
White-tailed Hare,	71

Mule Deer,	72
*Bighorn Sheep,	72

Birds

Turkey Vulture,	81
Golden Eagle,	81
Red-tailed Hawk,	81
California Quail,	83
Mountain Quail,	83
Great Horned Owl,	85
Allen's Hummingbird,	87
Gray Flycatcher,	93
Violet Green Swallow,	93
*Black-billed Magpie,	95
Bushtit,	97
House Wren,	97
Varied Thrush,	99
Orange-crowned Warbler,	103
Nashville Warbler,	103
MacGillivray's Warbler,	103
Wilson's Warbler,	105
Lazuli Bunting,	105
Towhees,	107
Fox Sparrow,	107

*Vesper Sparrow,	107
White-crowned Sparrow,	107
*Sage Sparrow,	107
*Black-throated Sparrow,	107
Lesser Goldfinch,	111

Reptiles

Western Fence Lizard,	115
Sagebrush Lizard,	115
Western Skink,	115
Southern Alligator Lizard,	115
Racer,	117
*Desert Striped Whipsnake,	117
California Whipsnake,	117
Common Kingsnake,	117
Calif. Mountain Kingsnake,	117
Sharp-tailed Snake,	117
N. Pacific Rattlesnake,	119

Amphibians

Calif. Slender Salamander,	123

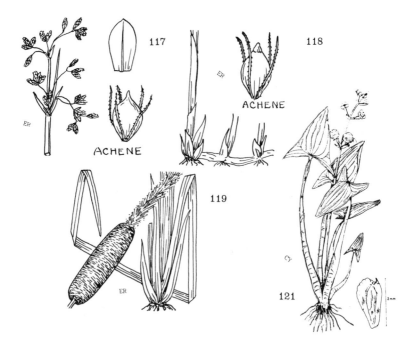

117

118

ACHENE

ACHENE

ER

119

121

ER

ER

FRESHWATER AREAS

Photo courtesy U.S. Forest Service.

Freshwater areas in the Sierra Nevada include streams, lakes, ponds, and swampy or boggy areas. Examples of some of the more common freshwater plants are given below. Sedges and rushes (page 35) are also found near fresh water.

117. **COMMON TULE**, *Scirpus acutus* var. *occidentalis*; Cyperaceae, Sedge fam. Perennial 5-13′ tall with cylindrical stems; forms dense stands in shallows.

118. **CREEPING SPIKERUSH**, *Eleocharis macrostachya*. 1½-4′ tall on long, creeping rhizomes; leaves form only basal sheaths; terminal spike up to 1″ long.

119.☆ **BROAD-LEAVED CATTAIL**, *Typha latifolia*; Typhaceae, Cattail fam. 4-9′ tall; 12-16 light green, swordlike leaves, exceeding stem in height.

120. **COMMON WATER-PLANTAIN**, *Alisma plantago-aquatica*; Alismataceae, Water-plantain fam. Plant 2-4′ tall; flowers in pyramidal, whorled panicles, generally white.

121.☆ **BROAD-LEAVED ARROWHEAD**, *Sagittaria* sp. Milky juice; leaves with sheaths on their bases; flowers in whorls of three. The tubers were often eaten by Indians, hence called Tule Potatoes.

122. **FLOATING-LEAVED PONDWEED**, *Potamogeton natans*; Potamogetonaceae, Pondweed fam. Most pondweeds have wide,

oblong to ovate floating leaves and narrow, linear submerged leaves; spike up to 2" long.

123. **WATER BUTTERCUP,** *Ranunculus aquatilus;* Ranunculaceae, Buttercup fam. Submerged stems generally 8-31" long; submerged leaves dissected into threadlike segments; above water leaves 3-lobed; flowers have light green sepals and 5 white petals, often yellow at base.

124. **MARE'S TAIL,** *Hippuris vulgaris;* Haloragaceae, Water-milfoil fam. Stem 1-2' tall, unbranched; linear to lanceolate leaves in whorls of approximately 7-10; tiny flowers have no petals. Usually found below 9000'.

125. **COMMON BLADDERWORT,** *Utricularia vulgaris;* Lentibulariaceae. Bladderwort fam. 1-3' long underwater stems shallowly float in quiet water; yellow flower brown-striped.

126. **YELLOW POND-LILY,** *Nuphar luteum* ssp. *polysepalum;* Nymphaeaceae, Waterlily fam. Leaves oblong to ovate, generally floating; flowers 2-2½" wide, yellow, sometimes red-tinged. North and central Sierra Nevada.

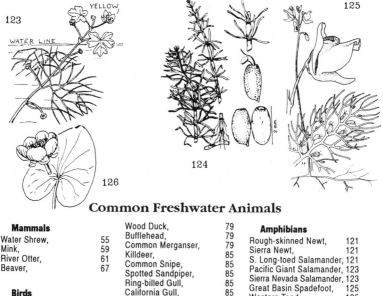

Common Freshwater Animals

Mammals	
Water Shrew,	55
Mink,	59
River Otter,	61
Beaver,	67

Birds	
Eared Grebe,	77
Pied-billed Grebe,	77
Great Blue Heron,	77
Canada Goose,	77
Mallard,	79
Green-winged Teal,	79
Northern Shoveler,	79

Wood Duck,	79
Bufflehead,	79
Common Merganser,	79
Killdeer,	85
Common Snipe,	85
Spotted Sandpiper,	85
Ring-billed Gull,	85
California Gull,	85
Belted Kingfisher,	89
American Dipper,	97

Reptiles	
Common Garter Snake,	119
Mountain Garter Snake,	119

Amphibians	
Rough-skinned Newt,	121
Sierra Newt,	121
S. Long-toed Salamander,	121
Pacific Giant Salamander,	123
Sierra Nevada Salamander,	123
Great Basin Spadefoot,	125
Western Toad,	125
Yosemite Toad,	125
Pacific Treefrog,	125
Foothill Yellow-legged Frog,	125
Red-legged Frog,	125
Northern Leopard Frog,	126

MAMMALS

Mammals are animals usually covered with hair or fur who give milk to their young. As most of the abundant species of mammals are entirely or mostly nocturnal, they are more likely to be seen at dusk or dawn, when it is hard to see them clearly. Although we have tried to show the pictures of animals in groups of comparative size, for various reasons this has not always proved possible. Therefore, when viewing a species illustration the reader should also note its accompanying measurements to correctly judge an animal's size. Besides this, as a basis for comparison, we use the four common mammals pictured below as examples, describing other animals shown as "raccoon +" (meaning larger than a raccoon), "rat -" (somewhat smaller than a rat), "mouse size" (same size as a mouse), and so on.

House Mouse 3 - 4"

Cat 15-18

House Rat 7 - 10"

Raccoon 18 - 28"

Wild mammals can best be approached by moving very slowly and quietly and wearing clothes colored like the surroundings. The illustrations of them are helpful for body shape and color pattern. Also study the descriptions and details about habits, noting particularly the habitats in which each mammal likes to live.

Order MARSUPIALIA: Marsupials. Prehensile tail; female with pocket.

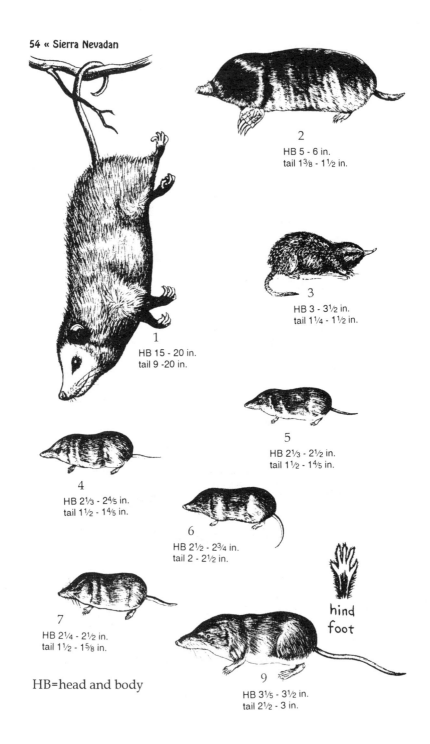

2
HB 5 - 6 in.
tail 1³⁄₈ - 1½ in.

3
HB 3 - 3½ in.
tail 1¼ - 1½ in.

1
HB 15 - 20 in.
tail 9 -20 in.

5
HB 2⅓ - 2½ in.
tail 1½ - 1⁴⁄₅ in.

4
HB 2⅓ - 2⁴⁄₅ in.
tail 1½ - 1⁴⁄₅ in.

6
HB 2½ - 2³⁄₄ in.
tail 2 - 2½ in.

hind foot

7
HB 2¼ - 2½ in.
tail 1½ - 1⁵⁄₈ in.

HB=head and body

9
HB 3⅕ - 3½ in.
tail 2½ - 3 in.

1. **VIRGINIA OPOSSUM**, *Didelphis virginiana*; Didelphidae, Opossum fam. **[str-wd.]** Cat + size. Coarse hair grayish; face white; hind foot with an opposable thumb; tail scaly; feigns death when in danger; unpleasant smell. Low elevations. Introduced in 1910.

Order INSECTIVORES. Numerous sharp teeth; small size.

2. **BROAD-FOOTED MOLE**, *Scapanus latimanus*; Talpidae, Mole fam. **[mead. str-wd.]** Mouse +. Hair light gray to coppery brown; the ears and eyes are not visible in thick fur; front feet as broad or broader than long; hunts underground invertebrates in soil. California's most common mole. (Illus. of mole hole p. 66.)

3. **SHREW-MOLE**, *Neurotrichus gibbsii*. **[mead. str-wd. sub-alp.]** Mouse size. Unlike the broad-footed mole, the front feet are longer than broad, the eyes are small but distinctly visible, and the tail is more haired; nose naked; fur dark or black; unlike other moles, this mole is often active above ground. In northern Sierra Nevada, south at least to Plumas Co.

4. **VAGRANT SHREW**, *Sorex vagrans*; Soricidae, Shrew fam. **[mead. str-wd. sub-alp.]** Mouse -. Color reddish brown in summer, nearly black in winter. Canine-like upper incisor teeth and long pointed nose distinguish shrews from mice. They usually dart about quickly, hunting insects and mice in thick vegetation and debris.

5. **ORNATE SHREW**, *Sorex ornatus*. **[str-wd. mead. sub-alp.]** Mouse -. Grayish brown above, lightly marked with silver-tipped hairs; paler below. West side of central to southern Sierra Nevada at low elevations. This is the common shrew of the Central Valley.

6. **TROWBRIDGE SHREW**, *Sorex trowbridgii*. **[mead. str-wd.]** Mouse-. Brown in summer, gray in winter; tail distinctly bicolored. Tulare Co. north at low elevations.

7. **MONTANE SHREW**, *Sorex monticolus*. **[mead. str-wd.]** Mouse -. Rusty brown above, ashy gray below, darker and grayer in winter; tail bicolored. From San Jacinto Mts. north, but not in north Coast Ranges.

8. **MOUNT LYELL SHREW**, *Sorex lyelli*. **[sub-alp.]** (Not illus.) Mouse-. 2¼-2½" long; tail 1½". Brown above, gray below. In grass or under stream-side willows. Rare above 7000' in the southern Sierra Nevada.

9. **WATER SHREW**, *Sorex palustris*. **[str-wd. water]** Mouse size. Hind feet large and lined with stiff white hairs to aid in swimming; upper parts black frosted with gray hairs; dense velvety fur is water-repellent; belly whitish tinged with brown. Tulare Co. north above 4000 feet.

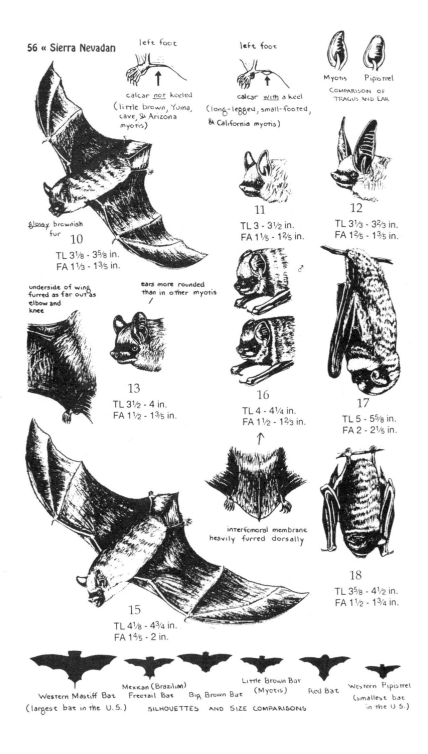

left foot

calcar *not* keeled
(little brown, Yuma, cave, & Arizona myotis)

left foot

calcar *with* a keel
(long-legged, small-footed, & California myotis)

Myotis   Pipistrel
COMPARISON OF
TRAGUS AND EAR

glossy brownish fur

10
TL 3⅛ - 3⅝ in.
FA 1⅓ - 1⅗ in.

11
TL 3 - 3½ in.
FA 1⅕ - 1⅖ in.

12
TL 3⅓ - 3⅔ in.
FA 1⅖ - 1⅗ in.

underside of wing furred as far out as elbow and knee

ears more rounded than in other myotis

13
TL 3½ - 4 in.
FA 1½ - 1⅗ in.

16
TL 4 - 4¼ in.
FA 1½ - 1⅔ in.

17
TL 5 - 5⅝ in.
FA 2 - 2⅕ in.

interfemoral membrane heavily furred dorsally

15
TL 4⅛ - 4¾ in.
FA 1⅘ - 2 in.

18
TL 3⅝ - 4½ in.
FA 1½ - 1¾ in.

Western Mastiff Bat
(largest bat in the U.S.)

Mexican (Brazilian)
Freetail Bat   Big Brown Bat

Little Brown Bat
(Myotis)   Red Bat

Western Pipistrel
(smallest bat in the U.S.)

SILHOUETTES AND SIZE COMPARISONS

Order CHIROPTERA: Bats. Flying Mammals. Some go south in winter, others hibernate in caves. Bats are important in keeping down the number of insects, such as mosquitoes. Their very rapid squeaks are often too high to be heard.

10. **LITTLE BROWN BAT,** *Myotis lucifugus;* Vespertilionidae, Common Bat fam. **[sub-alp. rocks, water, mead. chap. conif.]** Mouse size. Wingspread 10½"; hair dark brown with glossy tips. Like most bats, it catches insects in the dusk. This species appears in late summer in the high mts., usually flying high among tree tops. Found from Kern Co. north.

11. **CALIFORNIA BAT,** *Myotis californicus.* **[str-wd. mead. conif.]** Mouse size or less; wingspread 9½". Dark ears and face contrast with bright yellow-brown fur, with bases of hair dark; body paler below. Commoner at low elevations.

12. **LONG-EARED BAT,** *Myotis evotis.* **[conif. mead. rocks]** Mouse size. Golden brown and glossy fur; long black ears. Often seen darting back and forth between trees at dusk, generally in openings in ponderosa pine forest or lower red fir forest.

13. **HAIRY-WINGED BAT,** *Myotis volans.* **[conif. mead. str-wd. rocks]** Mouse size. Varies from dark brown to yellowish cinnamon above; wing membranes and ears blackish; ears short and rounded; fur on underarm extends out farther than on other Myotis.

14. **SMALL-FOOTED BAT,** *Myotis leibii.* (Not illus.) **[rocks, conif. mead. str-wd.]** Mouse size. Length 2⅓-3½", forearm 1⅕-1⅖". The black wing membranes and ears contrast with the light or golden brown fur; face blackish. The foot is relatively small, ¼-⅜"; flat-topped forehead. Found in middle mountain forests.

15. **BIG BROWN BAT,** *Eptesicus fuscus.* **[mead. str-wd. rocks]** Mouse +. The only large brown bat; glossy, bright brown fur. Wingspread 13". Nose, feet, wing membranes, and ears are blackish. Frequently enters buildings.

16. **RED BAT,** *Lasiurus borealis.* **[conif. str-wd.]** Mouse +. Fur reddish with white-tipped hairs; ears broad and rounded. Likes to roost in deciduous trees.

17. **HOARY BAT,** *Lasiurus cinereus.* **[conif.]** Mouse +. A very large bat with yellowish brown to chocolate-brown fur fringed with white (hoary); throat buffy. Ears black-bordered. Summer visitor, hanging in trees by day.

18. **SILVER-HAIRED BAT,** *Lasionycteris noctivagans.* **[conif. water, str-wd.]** Mouse size. Covered with dark brown to black hair tipped with white; ears short and rounded. Usually roosts in trees, but sometimes in buildings. From central Sierra Nevada north.

19
HB 5 - 6 ft.

20
HB 18 - 28 in.
tail 8 - 12 in.

21
HB 14 - 16 in.
tail 14 - 16 in.

22
male HB 16-17 in.
female HB 14 - 15 in.

23   HB 20 - 25 in
tail 13 - 15 in.

25
male HB 6 - 9 in.
female HB 5 - 7½ in.

24
male HB 13 - 17 in.
female HB 12 - 14 in.

26
male HB   9 · 11 in.
female HB   8 - 9 in.

27
HB 18 - 22 in.
tail 4 - 6 in.

Order CARNIVORA: Carnivorous Mammals.

19. **BLACK BEAR,** *Ursus americanus;* Ursidae, Bear fam. **[chap. conif. sub-alp. mead. str-wd.]** Weight 200-500 pounds. Cinnamon brown to black all over. An omnivorous feeder, eating everything from acorns to marmots.

20. **RACCOON,** *Procyon lotor;* Procyonidae, Coon fam. **[str-wd. conif.]** Fur grayish tipped with black; tail usually has six black rings, and face a black 'bandit' mask. Omnivorous feeder, often foraging near water.

21. **RING-TAILED CAT,** *Bassariscus astutus;* Bassariscidae, Cacomistle fam. **[chap. rocks, str-wd.]** Cat size. Hunts mice and rats; eats wild fruit. Light brownish above, whitish below; conspicuously ringed tail. A very shy animal of lower and middle altitudes.

22. **PINE MARTEN,** *Martes americana;* Mustelidae, Weasel fam. **[conif. sub-alp. mead. rocks]** Cat size. Fur yellowish brown shading to dark brown on tail and legs, pale buff patch on throat and chest. Spends much time in trees hunting squirrels. From Kern Co. north.

23. **FISHER,** *Martes pennanti.* **[conif. sub-alp.]** Raccoon size. Dark brown to nearly black; white tipped hairs give it a frosted appearance; slender, powerful body. Savage hunter of smaller life, even martens. From Kern Co. north.

24. **MINK,** *Mustela vison.* **[conif. water, str-wd.]** Cat size. All dark brown or black in color; white chin patch. Hunts fish, mice, rats, etc. Expert swimmer.

25. **SHORT-TAILED WEASEL** or **ERMINE,** *Mustela erminea.* **[conif. rocks, mead. sub-alp.]** Rat -. Tail about one-third of body and head length; brown above with white underparts and feet in summer; body turns white in winter; tail always with black tip. A fierce hunter of mice. From Tulare Co. north.

26. **LONG-TAILED WEASEL,** *Mustela frenata.* **[conif. sub-alp. rocks, mead.]** Rat size. Brown above with white underparts; very similar to above, but tail about one-half length of head and body together. Often distinguished by having a white or light colored facial mask. May become pure white in winter, but tail always has a black tip.

27. **BADGER,** *Taxidea taxus.* **[mead. sub-alp.]** Raccoon size. Silver or yellowish gray with median white stripe running from nose over top of head; black and white markings on head; powerful, very long front claws used for digging out ground squirrels.

29

HB 26 - 30 in.
tail 12 - 17 in.

28

HB 29 - 32 in.
tail 7 - 9 in.

31

HB 9 - 13½ in.
tail 4½ - 9 in.

30

HB 13 - 18 in.
tail 7 - 10 in.

34

HB 32 - 40 in.
tail 13 - 18 in.

32

HB 21 - 29 in.
tail 11 - 16 in.

35

HB 25 - 30 in.
tail 5 in.

36

HB 42 - 54 in.
tail 30 - 36 in.

33

HB 22 - 25 in.
tail 14 - 16 in.

28. **WOLVERINE**, *Gulo gulo*. [conif. sub-alp. mead. rocks] Raccoon +. Dark brown to blackish; looks like a small bear except tail is bushy; light stripe along each side; whitish on forehead. Feeds on marmots, squirrels, grubs, carrion, etc.; ferocious. From Kern Co. north to Placer Co.

29. **RIVER OTTER**, *Lutra canadensis*. [water, str-wd.] Raccoon size +. Dark brown with light colored throat; feet webbed and tail tapered for aid in catching fish.. From Tulare Co. north in or near water. Very playful, affectionate.

30. **STRIPED SKUNK**, *Mephitis mephitis*. [mead. str-wd. conif.] Cat size. Narrow white stripe up center of forehead; broad white patch divides in a V at about the shoulders; chiefly nocturnal; feeds on mice, insects, berries, etc.; throws a powerful scent. Lower elevations.

31. **SPOTTED SKUNK**, *Spilogale putorius*. [chap. str-wd. conif.] Rat +. Black with white stripes and spots; hunts insects and mice at night; stands on front feet to cast a protective scent from two glands when alarmed.

32. **GRAY FOX**, *Urocyon cinereoargenteus*; Canidae, Dog and Fox fam.[chap. rocks, str-wd.] Raccoon size. Grayish coat, with yellowish red along sides of neck, backs of ears, legs, and feet; bushy tail has a median black stripe down its entire length. Lower altitudes.

33. **RED FOX**, *Vulpes vulpes*. [conif. sub-alp. mead. rocks] Raccoon size. Usually with yellowish red fur, though three other phases are also found, the cross fox (with brown and gray), and the black and silver foxes. Distinguished by black ear tips, black feet, and white-tipped tail.

34. **COYOTE**, *Canis latrans*. [most habitats] Raccoon +. Gray, sandy or brown. Looks like a sheep dog, but with shy and slinky habits; tail held between legs when running; feeds on rodents and rabbits, but will eat almost anything.

35. **BOBCAT**, *Lynx rufus*; Felidae, Cat fam. [chap. mead. conif. rocks] Raccoon size. Gray or tawny, spotted with brown or black; short tail, tufted ears. Eats small animals.

36. **MOUNTAIN LION**, *Felis concolor*. [conif. sub-alp. mead. chap. rocks] Fur brownish gray to tawny; tip of long tail dark brown.

Order RODENTIA: Rodents. Animals with two upper and two lower large, front gnawing teeth.

37
HB 14 - 19 in.
tail 4½ - 9 in.

BR–YEL

YEL
or
TNY

38
HB 9 - 11 in.
tail 5 - 9 in.

39
HB 8 - 9 in.
tail 2⅕ - 3 in.

OR – BR

GY–BR

40
HB 6 - 8 in.
tail 2½ - 4¾ in.

BR

42
HB 4⅗ - 5⅓ in.
tail 2⅘ - 4 in.

37. YELLOW-BELLIED MARMOT, *Marmota flaviventris;* Sciuridae, Squirrel fam. **[mead. rocks, str-wd.]** Cat size. Thick body; tail bushy; fur yellowish grizzly brown with dull yellow belly; white patch between the eyes. Has shrill whistle when alarmed; takes sun baths on top of boulders; hibernates Aug. to Feb. Not in Coast Ranges.

38. CALIFORNIA GROUND SQUIRREL, *Spermophilus beecheyi.* **[rocks, mead. sub-alp.]** Rat +. A dark triangular band of fur extends from the head down and spreads over the middle back; shoulders and sides of head light gray to whitish; remainder of body gray to brownish with lighter flecks on hind parts; tail more bushy than most ground squirrels. Mainly at lower altitudes.

39. BELDING'S GROUND SQUIRREL, *Spermophilus beldingi.* **[mead.]** Rat size. Tail short in proportion to body; noticeable broad brownish band down the back contrasts with buffy or buffy white sides and belly. Often sits up very straight looking like a stake and so is called the "picket-pin." Has loud, clear whistle. From Fresno Co. north, but not in Coast Ranges.

40. GOLDEN-MANTLED GROUND SQUIRREL, *Spermophilus lateralis.* **[conif. sub-alp. mead. rocks]** Rat -. Looks like a large chipmunk, but has a thicker body; head and shoulders golden or copper-colored; one white stripe bordered by two long black stripes on each side of the back. From Kern Co. north.

41. SONOMA CHIPMUNK, *Tamias sonomae.* (Not illus.) **[chap. conif.]** Mouse +. Length 5-6"; tail 4-5"; reddish and gray; dorsal stripes may be indistinct; lateral light stripes dull gray or brownish; head stripes reddish brown; a black patch appears below the ear. Marin Co. north; not in Sierra Nevada mountains.

42. LODGEPOLE CHIPMUNK, *Tamias speciosus.* **[conif. sub-alp. rocks]** Mouse +. Distinct white and dark brown side stripes; black median dorsal stripe; top of head brown to gray with black facial stripes; most of body rusty and gray colored. This is the common chipmunk of the west side of the Sierra Nevada; not in Coast Ranges.

43. ALPINE CHIPMUNK, *Tamias alpinus.* **[rocks, alpine, sub-alp.]** (Not illus.) Mouse size or +. 4-4½". A small, pale, brownish yellow chipmunk without sharply contrasting stripes, dark side stripes are reddish or brownish, never blackish; underside of tail light orange-yellow. Rarely climbs trees. Found along crests of mts. Tulare to Tuolumne counties.

44
HB 3⅔ - 4½
tail 3 - 4½ in.

YEL-GY

45
HB 5-6 in.
tail 3½ - 4⅔ in.

JB

46
HB 4⅔ - 6½ in.
tail 3½ - 5⅗ in.

GY-BR

47
HB 5⅓ - 6½ in.
tail 3⅘ - 6 in.

48
HB 6 - 7 in.
tail 4¾ - 5 in.

49
HB 9 - 12 in.
tail 10 - 12 in.

GY

51
HB 5⅗ - 6 in.
tail 2 - 3 in.

50
HB 5½ - 6⅖ in.
tail 4⅓ - 5½ in.

JB

JB

44. **LEAST CHIPMUNK,** *Tamias minimus.* **[sage, conif.]** Mouse +. Rich gray and tawny with contrasting dark and light stripes; outer light stripes white; long tail lemon-yellow underneath. Mainly in sagebrush scrub north and east of the Sierra Nevada, but also in Jeffrey pine forests of east side of mts. This is the smallest of the chipmunks and the most wide-ranging geographically.

45. **LONG-EARED CHIPMUNK,** *Tamias quadrimaculatus.* **[conif. chap. mead. rocks, str-wd.]** Mouse +. This large, bright, reddish to tawny chipmunk is found only in the high Sierra; ears long, slender; body stripes indistinct; large conspicuous white patch behind each ear; stripe below ear black. From Madera Co. north to Plumas Co.

46. **MERRIAM'S CHIPMUNK,** *Tamias merriami* **[chap.]** Mouse +. Large dull-colored chipmunk; all the light stripes are gray, while the dark stripes are brown; tail long and flat with light yellow brown on the edges. From Tuolumne Co. south into northern Baja.

47. **SHADOW CHIPMUNK,** *Tamias senex.* **[conif.]** Mouse +. Large, dark brown chipmunk with often indistinct contrasting stripes; back side of ears bicolored, blackish in front, grayish behind; stripe below ear brown. From Fresno Co. north.

48. **CHICKAREE or DOUGLAS' SQUIRREL,** *Tamiasciurus douglasii.* **[conif. sub-alp.]** Rat -. Dark brown above with reddish tints; black stripe on each side contrasts markedly with yellowish white to rusty underparts (summer); tail hairs white-tipped; ears tufted in winter. The high-pitched, indignant chatter of the chickaree is familiar in the forest. Gathers pine and fir seeds to eat or store for winter. From Kern Co. north.

49. **WESTERN GRAY SQUIRREL,** *Sciurus griseus.* **[conif. str-wd.]** Rat +. Tail as long or longer than body, very fluffy; body light gray above, white below. Forages far more on the ground for seeds than does the Chickaree. At lower altitudes on west side of Sierra Nevada and in the Coast Ranges.

50. **NORTHERN FLYING SQUIRREL,** *Glaucomys sabrinus.* **[conif.]** Rat -. The soft and silky fur is brownish gray above and white below; the eyes are large and dark; the tail is flattened and the skin spread between the legs to help the squirrel glide from one tree to another. Comes out only at night.

51. **MOUNTAIN POCKET GOPHER,** *Thomomys monticola;* Geomyidae, Gopher fam. **[sub-alp. mead. alpine]** Rat -. Russet brown or tawny (summer), pale grayish brown (winter); nose and

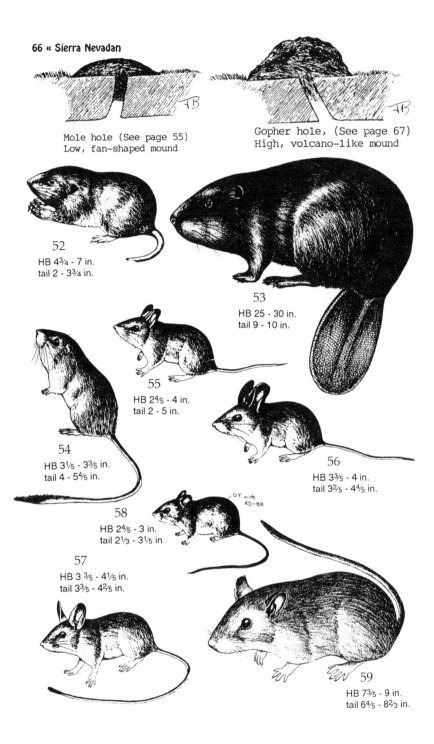

Mole hole (See page 55)
Low, fan-shaped mound

Gopher hole, (See page 67)
High, volcano-like mound

52
HB 4¾ - 7 in.
tail 2 - 3¾ in.

53
HB 25 - 30 in.
tail 9 - 10 in.

55
HB 2⅘ - 4 in.
tail 2 - 5 in.

54
HB 3⅕ - 3⅗ in.
tail 4 - 5⅘ in.

56
HB 3⅗ - 4 in.
tail 3⅖ - 4⅘ in.

58
HB 2⅘ - 3 in.
tail 2⅓ - 3⅕ in.

57
HB 3 ⅗ - 4⅕ in.
tail 3⅗ - 4⅖ in.

GY with
RD-BR

59
HB 7⅗ - 9 in.
tail 6⅘ - 8⅔ in.

patches behind pointed ears blackish; front feet strongly developed for digging; has outside cheek pouches; lives mainly underground, and feeds chiefly on tubers. Fresno Co. north at high altitudes.

52. **BOTTA'S POCKET GOPHER,** *Thomomys bottae*. **[mead.]** Rat -. Brown to grayish in color, but usually without strong black markings. Found in much lower meadows than *T. monticola*.

53. **BEAVER,** *Castor canadensis*; Castoridae, Beaver fam. **[water, str-wd.]**. Raccoon +. Fur rich brown; large, flat, hairless tail; cuts down trees with teeth; builds dams. From Tulare Co. north.

54. **CALIFORNIA POCKET MOUSE,** *Perognathus californicus*; Heteromyidae, Pocket Mice and Kangaroo Rat fam. **[chap.]** Has deep, fur-lined cheek pouches on either side of mouth on outside. Brownish gray above; yellowish white underparts and feet; tail bicolored and longer than head and body, crested on tip; black mark on nose; pale yellow-brown stripe along each side. El Dorado Co. south, low elevations.

55. **DEER MOUSE,** *Peromyscus maniculatus*; Cricetidae, Native Rat and Mice fam. **[all but water]** Tail furred and sharply bicolored. The soft, yellowish brown to grayish fur, large delicate ears, bright eyes, and clean habits are distinctive; feet and underparts white.

56. **PIÑON MOUSE,** *Peromyscus truei*. **[pin-jun. conif.]** Tail slightly shorter to slightly longer than head and body length, sharply bicolored; ears large, nearly 1 inch long; body brown to dark brown above, creamy white below. In piñon-juniper woodland, and higher in edge of Jeffrey pine forest.

57. **BRUSH MOUSE,** *Peromyscus boylei*. **[chap. rocks, conif. bldg.]** Tail usually a little longer than head and body; grayish brown to dark brown above, whitish below and on feet; ears fairly large. Found mainly at lower elevations. White-footed mice (*Peromyscus*) often come into abandoned cabins and make their nests out of mattress stuffings.

58. **WESTERN HARVEST MOUSE,** *Reithrodontomys megalotis*. **[mead. str-wd. rocks, chap.]** Distinguished by upper incisor teeth being grooved on their outer surfaces. Jumping mice also have such grooves, but their hind legs are large and powerful for jumping. Brown on back; light buff sides; grayish white tinged with buff below.

59. **DUSKY-FOOTED WOODRAT,** *Neotoma fuscipes*. **[str-wd. conif. chap.]** Grayish above, grayish to white below; hind feet

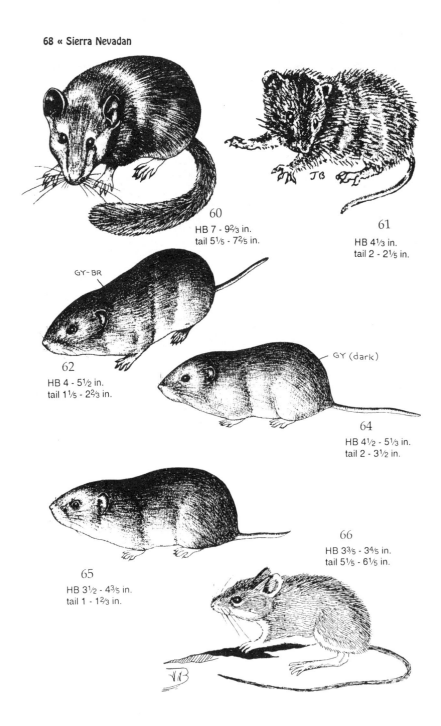

60
HB 7 - 9²/₃ in.
tail 5¹/₅ - 7²/₅ in.

61
HB 4¹/₃ in.
tail 2 - 2¹/₅ in.

GY-BR

62
HB 4 - 5¹/₂ in.
tail 1¹/₅ - 2²/₃ in.

GY (dark)

64
HB 4¹/₂ - 5¹/₃ in.
tail 2 - 3¹/₂ in.

66
HB 3³/₅ - 3⁴/₅ in.
tail 5¹/₅ - 6¹/₅ in.

65
HB 3¹/₂ - 4³/₅ in.
tail 1 - 1²/₃ in.

with dusky hairs on top. Builds large stick nests and is defensive of its home.

60. **BUSHY-TAILED WOODRAT**, *Neotoma cinerea*. **[rocks, mead. conif. sub-alp.]** Gray to brown with black hairs; long, bushy, squirrel-like tail distinctive. Woodrats are sometimes called "trade rats" because they pick up objects and drop others in their place.

61. **CALIFORNIA RED-BACKED VOLE**, *Clethrionomys californicus*. **[conif.]** Soft fur dark chestnut above, gradually grading into buffy gray on sides and below; often in old logs or stumps on forest floor. In north Coast Ranges.

62. **MOUNTAIN MEADOW VOLE**, *Microtus montanus*. **[mead. conif.]** Tail usually less than one third of body length; body thick, dark brown above, grayish below; blunt nose, tiny ears; often seen running along grassy runways. Found from Tulare Co. north, but not in northern Coast Ranges.

63. **OREGON VOLE**, or **CREEPING VOLE**, *Microtus oregoni*. (Not illus.) **[mead.]** 4-4½" long; tail 1¼-1⅝". Similar to the mt. vole, but with shorter fur. Found in northern Coast Ranges and Mt. Shasta area, then north.

64. **LONG-TAILED VOLE**, *Microtus longicaudus*. **[mead. sub-alp. str-wd.]** Fur dark gray washed with brown or blackish; tail bicolored; feet dirty white. Meadow voles usually have shallow tunnels underground where their nests are.

65. **MOUNTAIN PHENACOMYS** or **HEATHER VOLE**, *Phenacomys intermedius*. **[sub-alp. alpine]** Fur ashy gray, sometimes tinged with brown; belly silvery; feet white; does not build runways through the grass; often lives near patches of Sierra Heather. Found from Fresno Co. north; above 8000'.

66. **PACIFIC JUMPING MOUSE**, *Zapus trinotatus*; Zapodidae, Jumping Mouse fam. **[mead. conif. str-wd.]** The extremely long, scaly and untufted tail plus the long, strong hind legs are distinctive; sides orangish yellow; back dark brown to blackish with buffy tips; belly white. Travels by long leaps. In north Coast Ranges, but not in Sierra Nevada. **WESTERN JUMPING MOUSE**, *Zapus princeps*. **[mead. str-wd.]** Similar to *Z. trinotatus*, but not as brightly colored; ear bordered with yellow. Not in north Coast Ranges.

68
HB 7 - 10 in.
tail 5 - 8 in.

67
HB 3⅕ - 3⅖ in.
tail 2⅘ - 3⅘ in.

70
HB 18 - 22 in.
tail 7 - 9 in.

69
HB 12 - 17 in.
tail 1 - 1⅕ in.

72
HB 17 - 21 in.
ears 6 - 7 in.

71
HB 6⅕ - 8½ in.

73
HB 18 - 22 in.
ears 5 - 6 in.

74
HB 13 - 18 in.
ears 3½ - 4 in.

67. **HOUSE MOUSE,** *Mus musculus*; Muridae, Old World Mouse and Rat fam. **[bldg.]** Fur short, grayish brown; belly gray or buffy; tail scaly, naked. House pest.

68. **NORWAY RAT** or **HOUSE RAT,** *Rattus norvegicus*. **[bldg. str-wd.]** Brownish above, grayish below. A scavenger and pest.

69. **MOUNTAIN BEAVER,** *Aplodontia rufa*; Alplodontiidae, Mountain Beaver fam. **[str-wd. conif. sub-alp.]** Cat size. Fur dark brown and grizzled all over; tail hard to see; blunt nose and stocky body like the meadow vole, but much larger. It likes damp, thick vegetation, usually near streams, under which it makes extensive runways and burrows; not at all like the true beaver in habits. From Tulare Co. north in scattered colonies.

70. **PORCUPINE,** *Erethizon dorsatum*; Erethizontidae, Porcupine fam. **[conif. sub-alp. str-wd.]** Raccoon size. Distinctive sharp spines on back and tail; short legs, stout body; blackish tinged with yellowish brown. A slow-moving animal that climbs trees to eat the inner bark and small twigs; its quills protect it against enemies.

Order LAGOMORPHA: Conies and Hares. Four large, front gnawing teeth in upper jaw instead of only two as in rodents.

71. **PIKA** or **CONY,** *Ochotona princeps*; Ochotonidae, Pika fam. **[rocks, mead. sub-alp.]** Rat -. Ears round with white edges; no visible tail; soft fur grayish buff to dark brown; looks like a round-eared, short-legged rabbit or a guinea pig; gives a queer, ventriloquial whistled "eeenk-eeenk!" when alarmed. Eats green plants and dries them in piles on rockslides before storing them under the loose rocks for food in the winter. Found from Tulare Co. north, but not in Coast Ranges.

72. **BLACK-TAILED HARE** or **JACKRABBIT,** *Lepus californicus*; Leporidae, Rabbit and Hare fam. **[chap. sage, mead.]** Cat +. Color grayish brown; tail black above; large ears are black tipped; uses its powerful hind legs to escape enemies by running. Lower altitudes.

73. **WHITE-TAILED HARE,** *Lepus townsendii*. **[mead. sage pinjun.]** Cat +. Ears 5-6". In summer it is brownish gray with a white tail; in winter all white or pale gray except for its black-tipped ears. Tulare County north; not in Coast Ranges.

74. **SNOWSHOE RABBIT,** *Lepus americanus*. **[str-wd. conif. sub-alp.]** Cat size. Ears about 3". Very large, hairy feet, 4½-5¾" long, to aid in walking on snow; color brown in summer, white in winter. From Tuolumne Co. north, especially among willows.

Order ARTIODACTYLA: Hoofed Mammals.

75. **MULE DEER**, *Odocoileus hemionus*; Cervidae, Deer fam. **[conif. sub-alp. mead. chap. str-wd.]** Yellowish or reddish brown in summer, gray-brown in winter; tail either black-tipped or black on top; bucks grow a new set of horns each year; fawns dark-spotted on golden brown. The deer of the n.w. Sierra and northern Coast Ranges called the California Black-tailed Deer is now considered a subspecies of the Mule Deer.

76. **ELK**, *Cervus elaphus*. **[mead. conif.]** Bull has very large antlers, and makes a high-pitched bugling call during the mating season; dark brown mane; pale yellowish rump; back and sides grayish or reddish brown. Scattered in n.w. and near Owens Valley.

77. **BIGHORN SHEEP**, *Ovis canadensis*; Bovidae, Sheep and Ox fam. **[rocks, mead. sage, pin-jun.]** Brown to grayish brown with white rump; large, curved horns in males. Expert rock climbers. They move in small bands, mainly on east side of Sierra Nevada near Owens Valley.

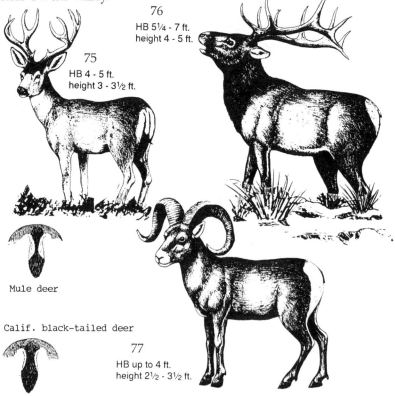

76
HB 5¼ - 7 ft.
height 4 - 5 ft.

75
HB 4 - 5 ft.
height 3 - 3½ ft.

Mule deer

Calif. black-tailed deer

77
HB up to 4 ft.
height 2½ - 3½ ft.

TRACKS

about 2¼"

GRAY SQUIRREL

about 1⅝"

CHIPMUNK

LEFT HIND FOOT

LEFT FORE FOOT

about 2¾" JACK RABBIT

← 7-12'
to next print →

about 6"

← 1 to 10' to
next print →

SNOWSHOE HARE

about 4¼"

RIGHT HIND FOOT

RIGHT FORE FOOT

RACCOON

RIGHT HIND FOOT

RIGHT FORE FOOT

about 2"

OPOSSUM

about 2"

LEFT FORE FOOT

LEFT HIND FOOT

MARTEN

about 3"

LEFT FORE FOOT

LEFT HIND FOOT

FISHER

about 2½"

HIND FOOT

COYOTE

about 1¾"

HIND FOOT

RED FOX

RIGHT HIND FOOT

RIGHT FORE FOOT

about 2"

MINK

TRACKS

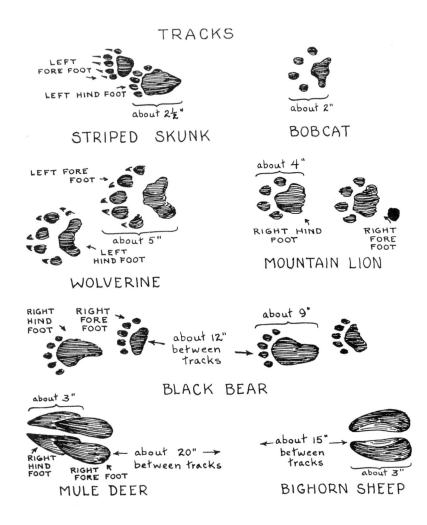

LEFT FORE FOOT
LEFT HIND FOOT
about 2½"
STRIPED SKUNK

about 2"
BOBCAT

LEFT FORE FOOT →
about 5"
LEFT HIND FOOT
WOLVERINE

about 4"
RIGHT HIND FOOT
RIGHT FORE FOOT
MOUNTAIN LION

RIGHT HIND FOOT
RIGHT FORE FOOT
about 12" between tracks
about 9"
BLACK BEAR

about 3"
RIGHT HIND FOOT
RIGHT FORE FOOT
about 20" between tracks
MULE DEER

about 15" between tracks
about 3"
BIGHORN SHEEP

BIRDS

Because birds have wings, they can move easily from one region to another. Some are resident all year, some are winter visitors, others spring and fall migrants. Unless otherwise described, a bird is assumed to be an all year resident. The bird descriptions that follow cover only the most important characteristics. But every bird is pictured and sometimes additional information is given if it is particularly useful in identification. Usually the male is pictured since the male is easier to identify, and the female is often present with the male. Note particularly types of bills and feet (page 76), sizes and shapes of bodies, manners of feeding and flying (page 112), and behaving. Note also the kind of habitat where each bird likes to live.

American robin
9-11"

common
house sparrow
6-6½"

yellow warbler
4½-5"

band-tailed pigeon
14-16"

common crow
17-21"

Comparitive size guide

Types of Bills

Nuthatch	*Woodpecker*	*Swallow*
Finch	*Hawk*	*Duck*
Shorebird	*Heron*	*Merganser*

Since the pictures of the birds on the pages that follow are often not in true proportion to each other in terms of real size, the reader is reminded to refer to each bird's measurements. Additionally, in the bird descriptions, sizes are given in comparison to the five familiar birds in the comparative size guide on the previous page. A "–" after the name means "a little smaller" than the example, and a "+" means "a little larger." Thus, crow + means a little larger than a crow.

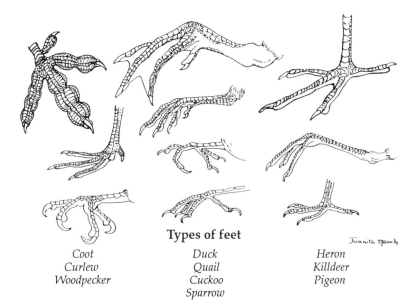

Types of feet

Coot	*Duck*	*Heron*
Curlew	*Quail*	*Killdeer*
Woodpecker	*Cuckoo*	*Pigeon*
	Sparrow	

Order PODICIPEDIFORMES, Family Podicipedidae: Grebes.

Grebes differ from ducks by their sharp bill, narrow head and neck (always held erect), labored flight, swift diving, and tailless appearance. They catch fish, frogs, etc.

1. **EARED GREBE,** *Podiceps nigricollis.* **[water]** Pigeon -. In summer the black crest on head, golden ear tufts and completely black neck and head are distinctive; in winter gray cheeks and white throat. Breeds mostly on lakes east of the Sierra Crest.

2. **PIED-BILLED GREBE,** *Podilymbus podiceps.* **[water]** Pigeon -. Drab brown grebe with thick, round bill; black bill ring and throat in summer; white throat and bill in winter. Occasional in spring and summer on lakes of lower elevations.

Order CICONIFORMES, Family Ardeidae: Herons and Bitterns.

3. **GREAT BLUE HERON,** *Ardea herodia.* **[water, str-wd. mead.].** About 4' high with a wingspread of 7'. Bluish gray with white head and neck and long legs. Herons are very large birds, curving neck in S-shape when flying. Summer visitor at higher elevations, all year around at lower elevations.

Order ANSERIFORMES, Family Anatidae: Ducks, Geese, Swans.

A. Geese. Larger than ducks; feed mainly on grasses and grain.

4. **CANADA GOOSE,** *Branta canadensis.* **[water, mead.]** Body gray-brown, head and neck black, with white chin band. Flock in V-shaped flight with hoarse, honking cries. Summer visitor primarily in the northeast part of the state, though it winters throughout the Sierra Nevada wildlife region.

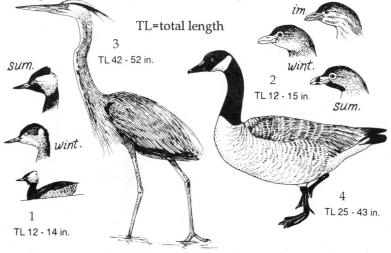

TL=total length

sum.

3
TL 42 - 52 in.

im.

wint.

2
TL 12 - 15 in.

sum.

wint.

1
TL 12 - 14 in.

4
TL 25 - 43 in.

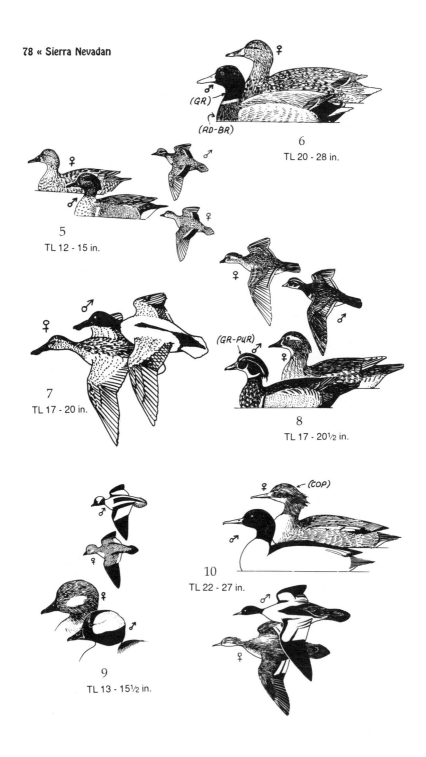

(GR)

(RD-BR)

6
TL 20 - 28 in.

♀

♂

♂

♀

5
TL 12 - 15 in.

♂

♀

7
TL 17 - 20 in.

♀

♂

(GR-PUR) ♂

♀

8
TL 17 - 20½ in.

♂

♀

♀

(COP)

♂

10
TL 22 - 27 in.

♀

♂

♀

9
TL 13 - 15½ in.

B. Surface-feeding Ducks. These obtain food by dabbling and tipping up the body instead of diving for it. In rising from the water, they spring straight up as shown.

5. **GREEN-WINGED TEAL**, *Anas crecca*. **[Water]** Pigeon -. A very small duck with a deep green speculum (wing patch); head of male reddish brown, with green stripe through eye; the back is gray; female duller, speckled light and dark brown. Occasional fall or spring migrant, mainly north of Lake Tahoe.

6. **MALLARD**, *Anas platyrhynchos*. **[water]** Crow +. Male has a glossy green head and neck, reddish brown breast, and white ring on its neck; female mottled brown with orange bill. All year around except at higher elevations when freezing temperatures drive them downslope.

7. **NORTHERN SHOVELER**, *Anas clypeata*. **[water]** Crow size. Bill is shovel-shaped; male head and neck dark green, sides and belly chestnut, lower neck and breast white; female mottled brown. Occasional in fall and winter on lower lakes and ponds.

8. **WOOD DUCK**, *Aix sponsa*. **[water, str-wd.]** Crow size. Male beautifully colored with a long downswept crest; head metallic green and purple with white stripes and throat; breast chestnut; back black; bright red area at base of bill. Female grayish brown with a smaller crest, white patch around the eye. Both sexes have nasal *"krrr-eeek"* call; male has a clear, rising mellow whistle. Low elevation.

C. Diving Ducks. These ducks dive deeply for food. In rising from the water they patter along the surface, as illustrated.

9. **BUFFLEHEAD**, *Bucephala albeola*. **[water]** Pigeon size. Male mostly white with a black back; puffy head has a large, bonnetlike white patch. Female much darker with a white mark on cheeks. Lower elevations in winter; forested mountain lakes north of Mt. Lassen in summer for breeding.

10. **COMMON MERGANSER**, *Mergus merganser*. **[water]** Crow +. Bright orange bill is narrow and saw-edged for fishing underwater; male black and white with a shiny green-black head (brown in the non-breeding season); female grayish with a reddish brown head and crest. A line of mergansers often fly over the

W=wingspan

GY (immature)
R (adult)

immature

adult

11
TL 26 - 32 in.
W 70 - 74 in.

12
TL 20 - 26 in.
W 42 - 48 in.

15
TL 30 - 41 in.
W 76 - 92 in.

13
TL 14 - 20 in.
W 27 - 36 in.

GY - BR

OR - RED

14
TL 10 - 14 in.
W 18 - 21½ in.

16
TL 19 - 25 in.
W 48 - 54 in.

water so low as to almost touch it. From Tulare Co. north; they usually nest along forested lakes and streams below 6000′; some small flocks occasionally stay through the winter.

Order FALCONIFORMES: Vultures, Hawks, etc.

(1) Family Cathartidae: Vultures. Feed on dead animals.

11. **TURKEY VULTURE,** *Cathartes aura.* [**sub-alp. chap. mead. rocks, conif.**] Wingspread 6′; naked red head (immature birds have dark gray heads); general color blackish with two-toned wings. Most are spring and summer visitors to the foothills, migrating south for the fall and winter..

(2) Family Accipitridae: Hawks and Eagles.

A. Short-winged Hawks (also called Bird Hawks). Distinctive long tails and broad short wings, which beat very rapidly.

12. **NORTHERN GOSHAWK,** *Accipiter gentilis.* [**conif. sub-alp. mead.**] Crow +. This is the biggest of the three bird hawks; upper parts uniformly gray; underparts light gray, finely barred; white stripe across the eye; immature bird is brown above with brown-streaked creamy underparts. From Kern Co. north.

13. **COOPER'S HAWK,** *Accipiter cooperii.* [**conif. sub-alp. str-wd.**] Crow -. Tail rounded; male bluish gray above, reddish brown, barred underneath; female larger and brownish blue above. Fiercely dives at birds through foliage.

14. **SHARP-SHINNED HAWK,** *Accipiter striatus.* [**conif. mead.**] Pigeon -. Looks like the Cooper's Hawk, but is smaller and has a square, often notched, tail. Nests in forests and edges of woods between ponderosa pine and red fir zones.

B. Buzzard Hawks and Eagles. These birds have very long, broad wings and short, usually broad tails.

15. **GOLDEN EAGLE,** *Aquila chrysaetos.* [**chap. rocks, conif. sub-alp. mead.**] Dark brown overall; seen closer up, it shows golden tinged hind neck. Soars high in sky, but drops like a meteor on prey; its greater size and longer wingspan distinguish it from the large buteos.

16. **RED-TAILED HAWK,** *Buteo jamaicensis.* [**rocks, str-wd. conif. sub-alp. mead.**] Crow +. Tail bright red above; thickest body and shortest tail of all the Buteo hawks. Cry, a shrill *keeer-r-r!* Like most hawks of this type, it does much good by holding the rodent and rabbit population in check.

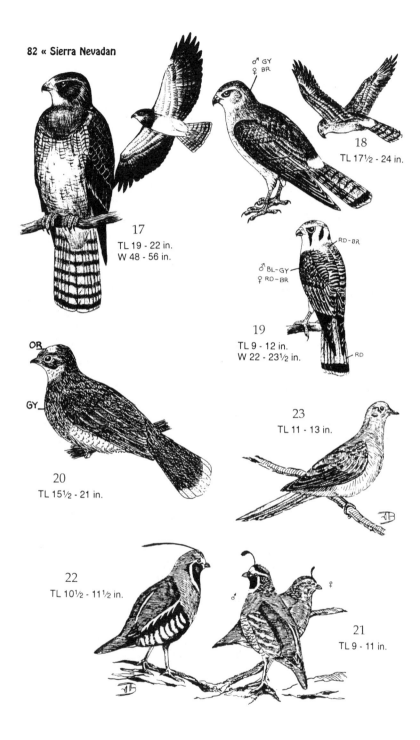

♂ GY
♀ BR

18
TL 17½ - 24 in.

17
TL 19 - 22 in.
W 48 - 56 in.

RD-BR

♂ BL-GY
♀ RD-BR

19
TL 9 - 12 in.
W 22 - 23½ in.

RD

OR

GY

23
TL 11 - 13 in.

20
TL 15½ - 21 in.

22
TL 10½ - 11½ in.

♂

♀

21
TL 9 - 11 in.

17. **SWAINSON'S HAWK**, *Buteo swainsoni*. **[mead.]** Crow size. In typical adults whitish wing linings contrast with dark flight feathers; dark, bib-like band below the throat; grayish tail narrowly banded. Has dark to light color phases. Lower elevations.

C. Harriers. Have long wings and long tails. They harry small animals by chevying them back and forth until they tire.

18. **NORTHERN HARRIER** or **MARSH HAWK**, *Circus cyaneus*. **[lower mead.]** Crow size. Male pale gray above, white beneath with reddish spotting; female brown above, streaked with brown on breast; both sexes display white rump. Immature bird is cinnamon below. Shrill *kek-kek-kek* cry.

(3) Family Falconidae: Falcons. Long, pointed wings; all have fairly long tails.

19. **SPARROW HAWK** or **AMERICAN KESTREL**, *Falco sparverius*. **[mead. rocks, conif. sub-alp. str-wd.]** Robin size. Male has blue-gray wings; both male and female reddish brown crown, back and tail; both with distinctive black and white facial markings; often hovers over fields watching for grasshoppers and mice.

Order GALLIFORMES, Family Phasianidae: Grouse, Quail, Partridges, etc.

20. **BLUE GROUSE**, *Dendragapus obscurus*. **[conif. sub-alp.]** Crow - to crow size. Male dusky gray or bluish gray overall with an orange comb above each eye; female and immature are mottled brown; tip of long rounded tail banded with light gray. From late April to early July males proclaim territory with a series of low, booming hoots.

21. **CALIFORNIA QUAIL**, *Callipepla californica*. **[chap. mead.]** Robin size. Forward curved black plume; males have black and white face and throat pattern. Rarely above foothill woodlands.

22. **MOUNTAIN QUAIL**, *Oreortyx pictus*. **[chap. conif. mead.]** Robin +. Gray and brown body, chestnut throat patch and distinct white markings on chestnut flanks; long, straight, black plume on head. Female similar but duller. Loud, resonant *t-woook!* or *woook* cry; *cut-cut* of female.

Order COLUMBIFORMES, Family Columbidae: Pigeons and Doves.

23. **MOURNING DOVE**, *Zenaida macroura*. **[conif. mead. str-wd.]** Pigeon-. Light brownish gray above; wings darker with black spots; tail long and pointed, edge showing white in flight; eye ring light blue; neck shield light, iridescent violet. Male has soft *coo-ooh, coo, coo-coo*. Found at lower elevations.

24
TL 14 - 16 in.

25
TL 9 - 11 in.

26
TL 10½ - 11½ in.

(wint.)

(sum)

27
TL 7½ - 8 in.

28
TL 20 - 23 in.

YEL

1st year

29
TL 18 - 21 in.

30
TL 18 - 25 in.

31
TL 6 - 7 in.

24. **BAND-TAILED PIGEON,** *Columba fasciata.* **[conif.]** Stout body; back and wings dark gray; tail with pale gray terminal band; head and underparts purplish; bill yellow tipped with black; feet yellow; white crescent on back of neck of adult; an owl-like *whoo-ooo* call.

Order CHARADRIIFORMES: Sandpipers and relatives.

(1) Family Charadriidae: Plovers.

25. **KILLDEER,** *Charadrius vociferus.* **[mead. water]** Robin size. Grayish brown above; white below with two black breast bands; long tail mostly tan. Pretends injury to lure intruders away from nest site. Shrill *kill-dee* cry; often cries constantly in flight.

(2) Family Scolopacidae: Sandpipers.

26. **COMMON SNIPE,** *Gallinago gallinago.* **[mead. water]** Robin size. Brown mottled with buff stripes on back; head striped; lighter underneath; very long bill. Zig-zag flight and raspy cry.

27. **SPOTTED SANDPIPER,** *Actitis macularia.* **[water, str-wd.]** Robin -. Olive brown above, with white line over the eye; belly and breast spotted with black in the summer (no spots in fall and winter); teeters up and down when walking. Summer along streams and lakes, but some remain all year.

(3) Family Laridae: Gulls.

28. **CALIFORNIA GULL,** *Larus californicus.* **[water]** Crow +. Adult with gray mantle; head and remainder of body white; eyes dark brown with reddish ring; red and black spot on lower mandible. Immature bird brownish overall, has black bar on end of bill. From Yosemite north on lakes.

29. **RING-BILLED GULL,** *Larus delawarensis.* **[water]** Crow size. Similar to the California Gull but smaller and mantle is lighter gray; a complete black ring encircles the bill in adults and eyes are yellow. Most common in northern Sierra Nevada.

Order STRIGIFORMES, Family Strigidae: Owls.

30. **GREAT HORNED OWL,** *Bubo virginianus.* **[rocks, str-wd. conif. sub-alp. mead. chap.]** Crow +. The only large owl with "horns," or ear tufts, the horns far apart; mottled gray-brown above; underparts light brown with dark bars; white throat bib conspicuous. Call is a deep *whoo-hoo-hoo!* Feeds on rodents and rabbits.

31. **FLAMMULATED OWL,** *Otus flammeolus.* **[conif.]** Sparrow +. Grayish brown above and light below, with white and rust colored markings; dark rather than yellow eyes; ear tufts inconspicuous; a mellow hoot is repeated steadily; very secretive. Migrates south for the winter.

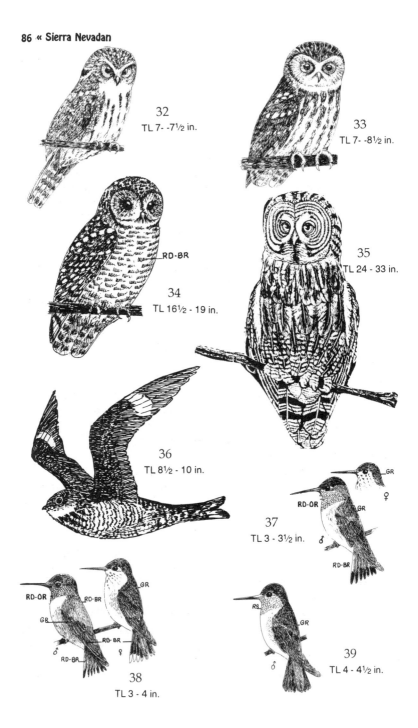

32
TL 7- -7½ in.

33
TL 7- -8½ in.

RD-BR

34
TL 16½ - 19 in.

35
TL 24 - 33 in.

36
TL 8½ - 10 in.

37
TL 3 - 3½ in.

RD-OR
GR
♀
GR
RD-BR
♂
RD-BR

38
TL 3 - 4 in.

RD-OR
RD-BR
GR
GR
RD-BR
♂
RD-BR
♀

39
TL 4 - 4½ in.

RD
GR
♂

32. **NORTHERN PYGMY OWL**, *Glaucidium gnoma*. [conif.] Sparrow +. Brown to gray-brown with fine buff spotting above; conspicuously brown-streaked buff below; tail long with white bars; no ear-tufts; two black patches on back of neck resemble eyes. Call, a single, mellow, whistled note repeated rapidly. Unlike most owls, pygmy owls often hunt by day.

33. **NORTHERN SAW-WHET OWL**, *Aegolius acadicus*. [conif. sub-alp.] Robin-. Brown, spotted with white above; no ear-tufts; wide, fluffy brown stripes in front; open wing shows white spots in rows. Remarkably tame if approached.

34. **SPOTTED OWL**, *Strix occidentalis*. [conif.] Crow size. Head round and without ear tufts; dark brown above but spotted white; light colored and heavily brown barred below; eyes large and dark brown instead of the usual yellow. Call high-pitched hoots, like barking of a small dog.

35. **GREAT GRAY OWL**, *Strix nebulosa*. [conif. sub-alp.] This is the largest of our owls; wingspread of 5'. Mottled gray and brown above; light gray breast streaked with dark brown; large, pale gray facial disk patterned in concentric circles; eyes and bill yellow. Deep, echoing *"hooo!"* Tame if approached; often hunts by day.

Order CAPRIMULGIFORMES, Family Caprimulgidae: Nighthawks.

36. **COMMON NIGHTHAWK**, *Chordeiles minor*. [mead. conif. sub-alp.] Robin size. Mottled brownish color; white bar across the pointed wing; male has white bar on notched tail and a white throat patch; flight swift and erratic, high in the air, in search of insects. Summer visitor.

Order APODIFORMES: Hummingbirds and Swifts.

(1) Family Trochilidae: Hummingbirds.

37. **ALLEN'S HUMMINGBIRD**, *Selasphorus sasin*. [mead. chap.] Warbler -. Male with bright orange-red throat, green crown and back, reddish brown rump, tail and sides; female duller in color, mostly olive green. Male makes U-shaped dive during courtship. Occasional in late summer at lower elevations.

38. **RUFOUS HUMMINGBIRD**, *Selasphorus rufus*. [mead.] Warbler -. Like Allen's Hummingbird having bright orange-red throat and reddish brown (rufous) rump, tail, and sides, but the back is rufous as well. Migrant in high meadows.

39. **BROAD-TAILED HUMMINGBIRD**, *Selasphorus platycercus*. [conif. pin-jun. sub-alp. mead.] Warbler -. Both sexes have metallic green upperparts, whitish underparts; male's gorget is rose colored. Male makes a unique, shrill trilling with wings in flight. East side of mts.

PUR-ULT

GR

♂

♀

40
TL 2¾ - 3¼ in.

41
TL 6 - 7 in.

42
TL 7 - 7½ in.

43
TL 11 - 14 in.

red

red

BR

GY

RED

♂

RED

RED

44
TL 12 - 14 in.

JB

45
TL 16 - 19½ in.

RED

46
TL 8 - 9½ in.

40. **CALLIOPE HUMMINGBIRD,** *Stellula calliope.* **[pin-jun. conif. mead.]** Warbler -. The tiniest bird of the mts. Male metallic green above, whitish below; gorget white with purple-violet rays. Females of these hummers look much alike. Migrates south by summer's end.

(2) Family Apodidae: Swifts. Swifts look like swallows, but have backward-curving, more pointed wings, are faster, and never rest on tree tops or telephone wires, but stay high in the air; the repeated short glide followed by the swift, twinkling beat of their stiff wings is characteristic.

41. **WHITE-THROATED SWIFT,** *Aeronautes saxatalis.* **[rocks]** Sparrow size. Contrasting black and white colors are distinctive. Summer visitor.

42. **BLACK SWIFT,** *Cypseloides niger.* **[rocks, mead.]** Sparrow +. All black except for a whitish patch on the forehead; tail notched and often fanned. Nests in deep, rocky canyons, often behind waterfalls. Summer visitor.

Order CORACIFORMES, Family Alcedinidae: Kingfishers.

43. **BELTED KINGFISHER,** *Ceryle alcyon.* **[water, rocks]** Pigeon -. Large head with bushy crest. Male blue-gray above with white underparts and blue-gray breast band. Female similar, but has reddish brown band below the blue-gray one; often hovers over water; has rattling cry; fishes for small fish and frogs.

Order PICIFORMES, Family Picidae: Woodpeckers. Strong bill used to dig in wood for grubs, insects, sap.

44. **NORTHERN FLICKER** (red-shafted var.), *Colaptes auratus.* **[conif. sub-alp. str-wd. mead.]** Pigeon -. Barred brown back, white rump, salmon-red under wings, black crescent on breast; female lacks the red mustache of the male. Likes ants.

45. **PILEATED WOODPECKER,** *Dryocopus pileatus.* **[conif. sub-alp.]** Crow size. Our largest woodpecker and the only crested one. Black overall; top of head and crest bright red; white underwing areas; white streak down side of neck; female has blackish forehead and lacks red on the mustache. In flight has a very loud excitable voice; one of its calls is an excited *kak-kak-kak*; loud hammering.

46. **ACORN WOODPECKER,** *Melanerpes formicivorus.* **[conif.]** Robin -. Red, white, and yellow patches on black head; back, wings and tail black; rump white; white patches on black wings. Pecks out holes in dead trees to store hoards of acorns. *Wik-up, wik-up* or *jay-cup, jay-cup* call.

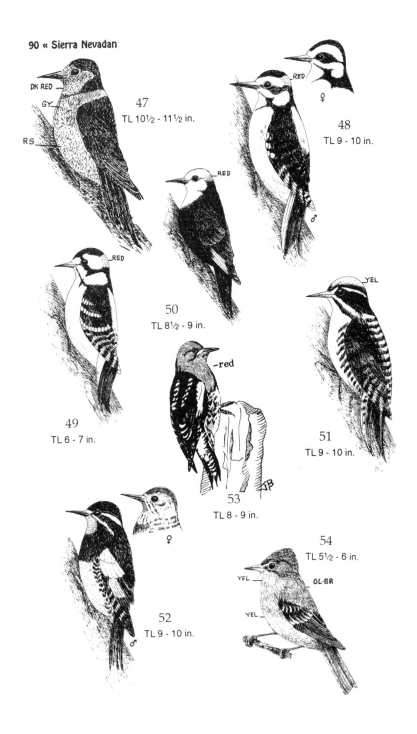

DK RED

GY

RS

47
TL 10½ - 11½ in.

RED

♀

48
TL 9 - 10 in.

♂

RED

50
TL 8½ - 9 in.

RED

49
TL 6 - 7 in.

-red

♀B

53
TL 8 - 9 in.

YEL

51
TL 9 - 10 in.

♀

52
TL 9 - 10 in.

♂

54
TL 5½ - 6 in.

YEL

OL-BR

YEL

47. **LEWIS' WOODPECKER** *Melanerpes lewis.* **[conif.]** Robin size. Iridescent greenish black above; collar and breast gray; extensive pinkish red belly; face dark red; *chrrr* call. Migrant.

48. **HAIRY WOODPECKER,** *Picoides villosus.* **[conif. sub-alp. str-wd.]** Robin size. White back and large bill are distinctive; red patch at base of crown is lacking in females; black wings spotted with white; black tail bordered with white. Note, a loud, sharp *peek!*

49. **DOWNY WOODPECKER,** *Picoides pubescens.* **[str-wd.]** Sparrow size. Looks like a small edition of the Hairy Woodpecker, except with shorter bill. A soft *pick* call; also has rapidly descending notes like a horse's whinny. Low altitudes.

50. **WHITE-HEADED WOODPECKER,** *Picoides albolarvatus.* **[conif. sub-alp.]** Robin size. The pure white head and black body are distinctive; male has red patch on nape; large white wing patches flash in flight. Call, a shrill *wick* or *chik*, often rapidly repeated.

51. **BLACK-BACKED WOODPECKER,** *Picoides arcticus.* **[sub-alp.]** Robin size. Black above, except for bright yellow patch on head of male; sides barred white and black below; underparts white; three toes. Tears bark. From Tulare Co. north; not in Coast Ranges.

52. **WILLIAMSON'S SAPSUCKER,** *Sphyrapicus thyroideus.* **[sub-alp. conif.]** Robin size. Male has black back, head, and breast; large white shoulder patch; a red throat and white face stripes. Female has a brown head and zebra-striped wings and sides. Both sexes have yellow belly and white rump. Nasal *churrr* call. Not in northern Coast Ranges.

53. **RED-BREASTED SAPSUCKER,** *Sphyrapicus ruber.* **[conif. str-wd. sub-alp.]** Robin -. Breast, neck, and head bright red; belly yellowish; back barred; long white wing patch and white rump. A secretive bird with a soft, nasal-sounding squeal. Summer visitor at higher elevations; some may winter in the foothills..

Order PASSERIFORMES: Perching Birds.

(1) Family Tyrannidae: Flycatchers. Flycatchers perch quietly on branch ends or telephone wires, suddenly circling out to snap up insects, then back to their perch again. The first four flycatchers listed below (*Empidonax* sp.) look very much alike, and are therefore best distinguished by habitat and song.

54. **PACIFIC-SLOPE FLYCATCHER,** *Empidonax difficilis.* **[str-wd. conif]** Sparrow -. Throat and belly yellowish; white eye ring and two pale wing bars; olive brown above. Song a squeaky

57
TL 5½ - 6 in.

56
TL 5 - 5½ in.

58
TL 6 - 6½ in.

59
TL 7 - 8 in.

60
TL 5 - 6 in.

61
TL 5 - 5½ in.

62
TL 7½ - 8½ in.

pseet-trip-seet! repeated over and over. Perches in shady places 10-30' high. Summer visitor.

55. **WILLOW FLYCATCHER**, *Empidonax traillii*. (Not illus.) **[str-wd.]**. Size 5½-6". Similar to above, but throat and belly pale. Explosive, clear-toned call of *weep-a-dee-yar!* or *fitz-bew!* Nests almost exclusively in dense willow thickets. Summer visitor.

56. **HAMMOND'S FLYCATCHER**, *Empidonax hammondii*. **[conif. sub-alp.]** Looks like above birds, but is olive-gray above; breast gray and belly pale yellow. Song a soft *seweep-tsurp-seep*, ending with a rising inflection. Usually perches 30' up or higher. Summer visitor.

57. **GRAY FLYCATCHER**, *Empidonax wrightii*. **[sage. pin-jun.]** Looks like above birds, but is gray above and whitish below; dips its tail like a phoebe. Song *chewip-cheep*, rising in tone. East side of mountains. Summer visitor.

58. **WESTERN WOOD PEWEE**, *Contopus sordidulus*. **[conif. sub-alp. str-wd.]** Sparrow size. Dark, olive-gray above, slightly lighter breast and sides; throat and belly light yellowish; no eyering; two whitish wing bars; call a rather harsh, nasal *pee-wee*. Summer visitor.

59. **OLIVE-SIDED FLYCATCHER**, *Nuttallornis borealis*. **[conif. sub-alp. str-wd.]** Sparrow +. Stout, large-headed flycatcher; dark brownish gray above; olive-gray of sides almost meets in center of dusky white chest; a white downy tuft may poke up from lower back. Loud, ringing call given from the tip-tops of conifers, *pip-pip-pip*. Song *whip-three-beers*. Summer visitor.

(2) Family Hirundinidae: Swallows. Wings are not as narrow and stiff as swifts, and their flight is graceful. They hunt insects by swooping low.

60. **TREE SWALLOW**, *Tachycineta bicolor*. **[conif. str-wd. mead.]** Sparrow -. Metallic blue-green-black above and white below. Immature birds gray-brown above. Always near water. Summer visitor.

61. **VIOLET-GREEN SWALLOW**, *Tachycineta thalassina*. **[conif. mead. chap.]** Warbler +. Shiny green and purple above, clear white below; white patches on rump and white extending above the eye distinguish it from the Tree Swallow. Summer visitor, nesting in cracks in cliffs, holes in trees, under eves, or in nest box.

62. **PURPLE MARTIN**, *Progne subis*. **[mead. conif.]** Robin -. Male is glossy blue-black over entire body; female whitish on the belly. Summer visitor, low elevations.

63
TL 7 - 8 in.

64
TL 12 - 13½ in.

65
TL 9 - 11¾ in.

66
TL 17 - 21 in.

67
TL 22- -27 in.

68
TL 12 - 13 in.

69
TL 17½ - 22 in.
tail 9½ - 12 in.

70
TL 5 - 5½ in.

71
TL 5 - 5¾ in.

(3) Family Alaudidae: Larks.

63. **HORNED LARK**, *Eremophila alpestris*. **[sub-alp. alpine]** Sparrow +. Black collar below yellow throat; black sideburns; brown on upperparts; two small black "horns"; walks instead of hops. From Nevada Co. to Lassen Co. in high mt. meadows.

(4) Family Corvidae: Crows and Jays.

64. **STELLER'S JAY**, *Cyanocitta stelleri*. **[conif. sub-alp. str-wd.]** Pigeon -. Front parts blackish; middle to rear parts deep blue; long crest. Has many harsh calls, often given rapidly; it copies the scream of the red-tailed hawk; also has a soft warble.

65. **PIÑON JAY**, *Gymnorhinus cyanocephalus*. **[pin-jun. conif.]** Robin size. Looks like a small gray-blue crow; white streaking on throat; no crest. Usually travels in leapfrogging flocks; high pitched *kaaa* with descending inflection has a mewing effect. Mainly on east side of Sierra Nevada.

66. **AMERICAN CROW**, *Corvus brachyrhynchos*. **[str-wd. mead. conif.]** All black; voice a harsh caw. Often travels in flocks; found at lower altitudes.

67. **COMMON RAVEN**, *Corvus corax*. **[rocks, sub-alp.]** Crow +. All black, but larger and stouter of body than the crow; throat feathers often appear rough or stick out; flies with wings straight out instead of bent up as does a crow. Voice a harsh croak; nests on cliffs or trees; feeds largely on carrion; often solitary.

68. **CLARK'S NUTCRACKER**, *Nucifraga columbiana*. **[sub-alp. conif. alpine]** Pigeon -. Light gray above; distinct white patch on trailing edge of each wing; black tail has white outer tail feathers. A harsh *kraa-aa!* call.

69. **BLACK-BILLED MAGPIE**, *Pica pica*. **[str-wd. sage, pin-jun. conif.]** Crow size. Large, slender bird with black and white colors and a long, sweeping tail; iridescent greenish on tail and wings. Staccato *kek-kek-kek* call; also nasal *maagh*. It is found on the lower edges of the middle mountain forests east of the Sierra Nevada Crest.

(5) Family Paridae: Chickadees and Titmice.

70. **PLAIN TITMOUSE**, *Parus inornatus*. **[conif. pin-jun.]** Warbler +. Small gray-backed bird with short crest; pale gray underparts. Harsh *chick-a-dee-dee* call, but soft, melodious song of *sweety-sweety-sweety*. Low elevations.

71. **MOUNTAIN CHICKADEE**, *Parus gambeli*. **[sub-alp. conif.]** Warbler +. White cheeks with line extending over the eye, black cap, black bib, and gray back; pale gray flanks. A very active

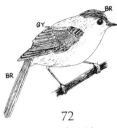

72
TL 3¾ - 4 in.

73
TL 5 - 6 in.

74
TL 4½ - 4¾ in.

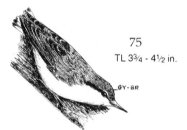

75
TL 3¾ - 4½ in.

76
TL 5 - 5¾ in.

77
TL 7 - 8½ in.

78
TL 4½ - 5 in.

insect hunter, sometimes hunting upside down. Song a clear whistled *fee-bee-bee*. The **CHESTNUT-BACKED CHICKADEE,** *Parus rufescens,* **[conif.]** with a chestnut back, is commoner in the north Coast Ranges, but found as far south as Yosemite. Many varying notes.

(6) Family Aegithalidae: Bushtits.

72. **BUSHTIT,** *Psaltriparus minimus.* **[chap. mead. str-wd.]** Warbler -. Tiny, gray-backed bird, usually moving in twittering flocks; its long tail is characteristic. Pacific Coast bushtits have a brown crown.

(7) Family Sittidae: Nuthatches.

Small short-tailed birds that climb on bark both head up and head down, hunting for insects in cracks.

73. **WHITE-BREASTED NUTHATCH,** *Sitta carolinensis.* **[conif. sub-alp.]** Sparrow -. Blue-gray back, black cap, white undersides. Low nasal *yank* call. Found from Kern Co. north.

74. **RED-BREASTED NUTHATCH,** *Sitta canadensis.* **[sub-alp. conif.]** Warbler size. Reddish brown instead of white below, blue-gray above; black stripe through eye. Both nuthatches have nasal, repetitive calls, but the red-breasted's is a higher, tin-horn sounding *yank-yank*.

75. **PYGMY NUTHATCH,** *Sitta pygmaea.* **[conif.]** Warbler -. Grayish brown cap comes down to the eye; back and wings bluish gray; creamy white below. A shrill, staccato *ki-dee, ki-dee;* also a piping *ket-ket-ket* call.

(8) Family Certhiidae: Creepers.

76. **BROWN CREEPER,** *Certhia americana.* **[conif. sub-alp.]** Warbler +. Mottled brown above, white below; whitish eyebrow patch; uses its stiff tail to brace itself in climbing; climbs upward in spiral, then flies to bottom of next tree and starts up again. Call a light *see-see-see*.

(9) Family Cinclidae: Dippers.

77. **AMERICAN DIPPER** or **WATER OUZEL,** *Cinclus mexicanus.* **[water]** Robin -. Often dips its slate gray, chunky body up and down. A land bird turned water bird, it swims and walks under water. Call a sharp *zeet*. Found near fast-flowing streams in the mountains.

(10) Family Troglodytidae: Wrens.

Tails tip up at a cocky angle; they flick them about in a characteristic way.

78. **HOUSE WREN,** *Troglodytes aedon.* **[str-wd. conif. sub-alp. chap. bldg.]** Warbler size. Gray-brown above with dark markings; has no white marks. Song gurgles and stutters, loud and then softer; many scolding notes. Summer visitor.

79
TL 4 - 4½ in.

80
TL 5½ - 5¾ in.

RD-BR

GY   RD-BR

81
TL 5 - 6 in.

DK GY

RD-BR

juv.

82
TL 9 - 11 in.

83
TL 9 - 10 in.

OR
OR

84
TL 6½ - 8 in.

RD-BR

BUFF

BR

BL

RD-BR

RD-BR

BL

BL

BR

♀

JMc

85
TL 6½ - 7¾ in.

86
TL 6 - 7 in.

87
TL 6½ - 8 in.

79. **WINTER WREN,** *Troglodytes troglodytes.* **[conif.]** Warbler -. A tiny, dark wren with a very short tail sharply cocked; heavily barred belly; often bobs its body up and down. From Tulare Co. north.

80. **CANYON WREN,** *Catherpes mexicanus.* **[rocks, sub-alp.]** Warbler +. Reddish brown color, especially on belly, but breast and throat white. Song sprays out liquid notes in a staccato, then long double notes, ending in a deep *twee-twee-twee!* Moves higher in summer, lower in winter.

81. **ROCK WREN,** *Salpinctes obsoletus.* **[rocks]** Sparrow -. These last 3 wrens all bob up and down. This is a large gray wren with buff-tipped outer tail feathers, a light streak over the eye, rusty rump, and fine streaking on pale breast.

(11) Family Muscicapidae, Subfamily Turdinae: Thrushes.

82. **AMERICAN ROBIN,** *Turdus migratorius.* **[conif. sub-alp. mead. alpine str-wd.]** Distinctively red or reddish brown breast; blackish head and tail; dark gray back, yellow bill. Song a clear caroling. Most move to the foothills during the winter.

83. **VARIED THRUSH,** *Ixoreus naevius.* **[chap. conif.]** Robin size and very similar to a robin, but with a dark collar across orange brown breast, orangish eye stripe, and orangish wing bars. Soft *tchook* cry; clear, breezy song. Winter visitor.

84. **HERMIT THRUSH,** *Catharus guttatus.* **[conif. sub-alp.]** Sparrow +. Brown back, spotted breast, and reddish brown tail and rump. A long and extremely beautiful song, flutelike in quality; *chuk* alarm note; a nasal *psee* and clear *kee* calls. Most Sierra breeders migrate to Mexico by late August but are replaced by other Hermit Thrushes from the north. These descend to the foothills for the winter.

85. **SWAINSON'S THRUSH,** *Catharus ustulatus.* **[str-wd.]** Sparrow +. Olive-brown to gray-brown above without any reddish brown on tail or rump as in *C. guttatus*; conspicuous buffy eye-ring and cheek; spotted breast. A soft, beautiful song, gradually rising in pitch. Summer at lower elevations.

86. **WESTERN BLUEBIRD,** *Sialia mexicana.* **[conif. sub-alp. mead. str-wd.]** Sparrow +. The only bird of this size with a blue tail, head, throat, and wings, while the breast and back are reddish. Female gray above with dull blue wings and tail. Common inhabitant of the border between woods and open areas.

87. **MOUNTAIN BLUEBIRD,** *Sialia currucoides.* **[conif. sub-alp. mead. alpine]** Sparrow +. Male turquoise blue all over except for the white belly. Female light gray-brown with blue on tail

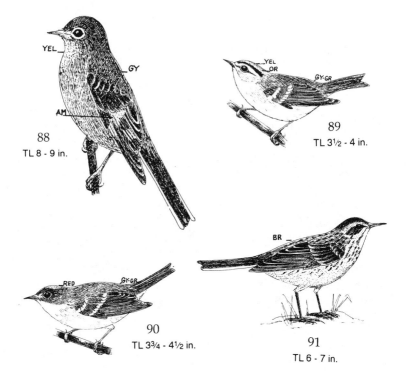

YEL

GY

AM

88
TL 8 - 9 in.

YEL
OR

GY-GR

89
TL 3½ - 4 in.

RED

GY-GR

90
TL 3¾ - 4½ in.

BR

91
TL 6 - 7 in.

GY

92
TL 4½ - 5½ in.

OL

GY

YEL

93
TL 5 - 6 in.

region and wings. A simple *cu-cu-cu* song. Most descend to lowlands for winter.

88. **TOWNSEND'S SOLITAIRE**, *Myadestes townsendi*. **[sub-alp. conif.]** Robin -. Gray colors, white eye ring, white-sided tail, and amber patch in center of wing are distinctive. A creaking *eesk* call; the male sometimes makes its long, warbling song while spiraling into the sky, then zig-zaging in steep pitches back to earth.

(12) Family Muscicapidae, Subfamily Sylviinae: Kinglets. Constantly flit among twigs and leaves.

89. **GOLDEN-CROWNED KINGLET**, *Regulus satrapa*. **[conif. sub-alp.]** Warbler -. Bright crown patch (orange in male, yellow in female); white stripe over eye; grayish green above, whitish below. Moves quickly. High, hisslike *see-see* note.

90. **RUBY-CROWNED KINGLET**, *Regulus calendula*. **[conif. sub-alp.]** Warbler -. Olive gray above; incomplete white eye ring; male with scarlet crown patch (usually concealed; erect when excited). A harsh *ji-dit* note, and a *chep-chep!* alarm note. Its song begins with 2-3 shrill notes, followed by a softer phrase, then *ti-dee-dee* repeated.

(13) Family Motacillidae: Pipits.

91. **WATER PIPIT**, *Anthus spinoletta*. **[sub-alp. mead. alpine]** Sparrow size, but with a slender bill and body; grayish brown above; light buff breast and sides streaked with brown. Characteristic teetering of tail and walks instead of hops. Found in high meadows in summer; they winter in the foothills.

(14) Family Vireonidae: Vireos. Vireos are often mistaken for small flycatchers or warblers, but they are duller colored than most warblers, more secretive and slower-moving. They do not take the upright position on a twig as do the flycatchers and they have more warbling songs; also vireo eye rings look like spectacles.

92. **WARBLING VIREO**, *Vireo gilvus*. **[str-wd. sub-alp.]** Warbler size. A dull, olive-gray bird with no distinguishing marks; has a long, slow, warbling song. Low, soft *chu-chuh* note. Summer visitor.

93. **SOLITARY VIREO**, *Vireo solitarius*. **[str-wd. conif.]** Sparrow -. Has white wing bars, large white spectacle-like eye rings, and a white throat and underparts; olive green back; gray crown; yellowish flanks. Short, variable, whistled song; *chee-wee* and *whee-wee* notes. Summer visitor.

94
TL 4½ - 5½ in.

95
TL 4 - 5 in.

96
TL 4½ - 5 in.

97
TL 5 - 6 in.

98
TL 4½ - 5 in.

99
TL 4½ - 5 in.

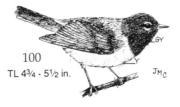

100
TL 4¾ - 5½ in.

(15) Family Emberizidae

A. Subfamily Parulinae: Wood Warblers. Usually brightly colored birds with thin, short bills; flit about restlessly among twigs.

94. **ORANGE-CROWNED WARBLER,** *Vermivora celata.* [**str-wd. chap. conif. sub-alp.**] The plainest looking of all the warblers; olive green above, yellowish green below; orange of crown seldom visible; no eye-ring. Song a simple trill, rising in pitch, then descending, repeated over and over. By July and August small flocks move up to the sub-alpine forests and brushy parts of alpine meadows. Most migrate south by October, but a few stay in the foothills through the winter.

95. **NASHVILLE WARBLER,** *Vermivora ruficapilla.* [**conif. chap.**] Olive-green upperparts, but underparts bright yellow; a white eye ring and a gray head; males have a dull reddish crown patch, which is seldom visible. Song a rapid series of 2-part notes, followed by a short trill of single notes: *seebit, seebit...tititititi.*Summer visitor from Tulare Co. north.

96. **YELLOW WARBLER,** *Dendroica petechia.* [**str-wd.**] The only overall yellow warbler; male with light reddish brown breast streaks. Light *see-see-wee-tee-see* song. Summer visitor.

97. **YELLOW-RUMPED WARBLER** (Audubon's), *Dendroica coronata.* [**sub-alp. conif.**] Male in spring is blue-gray above with dark streaks; black breast patch; throat, crown, and side patches yellow; Female in spring is brown with yellow patches not so vivid; both sexes are brownish above in winter. Often catches insects like the flycatcher by flying out from a twig to catch them. Song of *seet-see-tseet* notes, followed by a low-pitched trill. This is the only warbler that commonly resides year-round in the Sierra Nevada, moving to the foothills for the cold months.

98. **HERMIT WARBLER,** *Dendroica occidentalis.* [**conif. sub-alp.**] Male with yellow head and black chin and throat; gray back and wings; white wing bars and outer tail feathers; white belly and breast. Female has little or no black throat markings. Soft *chup* call; brisk, variable song. Summer visitor.

99. **BLACK-THROATED GRAY WARBLER,** *Dendroica nigrescens.* [**conif. pin-jun.**] Male gray with black and white striped face and black throat; yellow spot between bill and eyes; female with whitish throat. Variable, wheezy song starts with *zeedle, zeedle* or *zee-zee, zee-zee* in lazy swing.

100. **MACGILLIVRAY'S WARBLER,** *Oporornis tolmiei.* [**chap. conif. sub-alp. str-wd.**] Male has a slate gray hood covering head and neck, turning blackish on breast; olive green back and wings;

101
TL 4½ - 5 in.

102
TL 8 - 10 in.

103
TL 7 - 8½ in.

104
TL 6 - 7½ in.

105
TL 6½ - 7¾ in.

106
TL 5 - 5½ in.

yellow below; female with paler hood. Song a rolling chant of *seedle, seedle* or *sweeter, sweeter*. Call note a loud, sharp *tick* or *chip*. Very secretive, often jerks tail. Summer visitor.

101. **WILSON'S WARBLER,** *Wilsonia pusilla*. **[str-wd. chap. sub-alp.]** Olive green above and sharply yellow below; black eyes on yellow face; male has a black cap. Song begins with a rapid *chit-chit-chit*, getting louder and faster. Summer visitor.

B. Subfamily Icterinae: Blackbirds, Orioles, etc.

102. **BREWER'S BLACKBIRD,** *Euphagus cyanocephalus*. **[conif. mead. str-wd]** Robin size. Male mostly black with purplish and greenish tints; yellow eyes. Female brownish with eyes dark. Metallic *check* note; gurgling, wheezy song.

103. **BULLOCK'S ORIOLE** (Northern), *Iceterus galbula bullockii*. **[str-wd.]** Robin -. Male with black upper back and top of head, black on the throat, black line through eye; rest of head and body orange; wings dark with a broad white patch. Female olive gray above with yellow throat and chest, and white belly. Immature bird similar in appearance to female, but has black throat. Variable whistled song. Summer visitor.

C. Subfamily Thraupinae: Tanagers.

104. **WESTERN TANAGER,** *Piranga ludoviciana*. **[conif. sub-alp.]** Sparrow +. Male has bright red head, yellow body, black wings, back, and tail; female is yellow-green above, yellowish below. Both have two wing bars. Song like the robin's, except hoarser; call a dry *pit-r-ick*. Summer visitor.

D. Subfamily Cardinalinae: Buntings and Grosbeaks.

105. **BLACK-HEADED GROSBEAK,** *Pheucticus melanocephalus*. **[str-wd. conif.]** Sparrow +. Male has black head, dull orange-brown breast, collar and rump; black and white wings; a very thick bill. Female duller, mostly brown with dark streaks above, striped head. A sharp *eek* call; liquid, mellow song of lifting and falling notes. Summer visitor. Note: This is a distant relative of the grosbeaks of the Family Fringillidae.

106. **LAZULI BUNTING,** *Passerina amoena*. **[chap. str-wd]** Warbler +. Male bright blue above, except for black and white wings; cinnamon breast and sides; white belly. Female dull brown. Lively, shrill song, starting with 2 notes: *sweet-sweet* or *chew-chew*, etc.; a *tsik* alarm note. Male often sings throughout the day from high perches. Summer visitor.

BR

GY-BR

107
TL 8 - 10 in.

RE♂

108
TL 7 - 8½ in.

RD - BR

RD-BR

109
TL 6½ - 7 in.

RD-BR

110
TL 5 - 5¾ in.

111
TL 6½ - 7½ in.

112
TL 5½ - 6 in.

GY

115
TL 4¾ - 5½ in.

GY

113
TL 6 - 7 in.

114
TL 5 - 6 in.

E. Subfamily Emberizinae: Towhees, Sparrows and Juncos.

107. CALIFORNIA TOWHEE or **BROWN TOWHEE,** *Pipilo crissalis.* **[chap. pin-jun.]** Robin -. Earth brown above with buffy, lightly streaked throat; underparts lighter; rust colored under-tail coverts. Buzzy notes; metallic *chink.*

108. RUFOUS-SIDED TOWHEE, *Pipilo erythrophthalmus.* **[chap.]** Robin -. Black above; back and wings marked with white, rufous sides, white belly; eyes red. Female has the same pattern but is brown where the male is black. Loud *meow* like note; song is a loud, buzzing trill. Scratches noisily in dead leaves.

109. GREEN-TAILED TOWHEE, *Pipilo chlorurus.* **[chap.]** Sparrow +. Dull olive green above, gray breast; undersurface of wings yellow; reddish brown cap and white throat and belly Churring song; soft *mew* call. Summer visitor.

110. CHIPPING SPARROW, *Spizella passerina.* **[conif. sub-alp. mead.]** Plain gray breast; white line above the eye and black line across the eye; reddish brown cap. Song, a one-pitched chipping rattle; sweet *seep* call note. Summer visitor.

111. FOX SPARROW, *Passerella iliaca.* **[chap. str-wd.]** Large, stout sparrow; brownish gray with heavily-streaked breast and belly; reddish brown tail. Does much vigorous scratching under bushes. Song, one or more soft, sweet notes, followed by several short trills; metallic *sisp* note.

112. VESPER SPARROW, *Pooecetes gramineus.* **[mead. sage]** Flashes its white outer tail feathers in flight; chestnut patch on wing base; body brown above, dusky streaked. Song starts with 2 clear, flutelike notes, followed by 2 or 3 short, higher trills. Mainly east side of mountains, but also southern Sierra on the west side.

113. WHITE-CROWNED SPARROW, *Zonotrichia leucophrys.* **[mead. str-wd. chap.]** Crown white and black striped; cheek, neck and breast gray; belly whitish; back streaked gray and brown with white wing bars. Metallic *chink* or *pink* call; song starts with *saay-see-say,* followed by a trill.

114. SAGE SPARROW, *Amphispiza belli.* **[sage, chap.]** Color generally gray; distinctive black marks on sides of the throat and a single black spot on breast. Often flips long, square tail. A soft *kik-kik* call; a thin, high and jerky song. More common east of the mountains and in the southern Sierra as a summer visitor, though a darker race lives year-round from El Dorado to Mariposa Co.

115. BLACK-THROATED SPARROW, *Amphispiza bilineata.* **[sage]** Gray above; black throat and breast patch, white stripes on the face; whitish below. Immature bird lacks black bib. Cheerful

116
TL 5 - 6 in

117
TL 5½ - 6¼ in.

RD-BR

RD-BR

GY

YEL

YEL

YEL

♂

♀

118
TL 7 - 8½ in.

119
TL 5½ - 6 in.

ROSE

ROSE

♀

♂

BR

ROSE

BR

ROSE

BR

♀

RS

BR

121
TL 5¾ - 6¾ in.

♂

GY

BR

120
TL 6 - 6½ in.

RS

see-see'tsee or *sweet-sweet-wee* song, ending in a higher or lower trill. Generally found on the east side of the Sierra Nevada. Summer visitor.

116. **LINCOLN'S SPARROW**, *Melospiza lincolnii*. **[mead. sub-alp.]** Grayish brown above with dark brown markings; buffy breast finely streaked with brown; head partly gray. Soft to loud *tsee* or *tsep* call; gurgling song starts with low notes, rises and lowers rapidly, and ends softly. Often the song has pauses between phrases.

117. **OREGON JUNCO**, *Junco hyemalis oreganus*. **[sub-alp. conif. mead.]** Reddish brown back and black head; tail feathers flash white in flight; buff or reddish brown on sides; white belly. Female grayer on head and duller. Ringing, insectlike trill on the same pitch. Also twitters and makes clicking noises. Often in large flocks.

(16) Family Fringillidae: Finches. All have thick, strong bills for seed-cracking.

118. **EVENING GROSBEAK**, *Coccothraustes vespertina*. **[sub-alp. conif.]** Sparrow +. A stocky finch with a very large conical bill; male with dark yellow body; yellow patch on forehead and above eyes; head, neck, and breast dark olive brown. Female brownish gray with yellowish wash; both sexes have black wing and tail feathers heavily patched with white. A shrill, chirped *tseer-ip* call.

119. **PURPLE FINCH**, *Carpodacus purpureus*. **[conif. str-wd.]** Sparrow -. Male dull rose-red above except for brown-streaked mantle, and brown wings and tail. Female grayish brown with dark streaks on whitish underparts; noticeable dark patches on ear and jaw. Call in flight a sharp *pit!*. Song, a fast, lively warble. These are highly social birds and tend to feed in flocks. Summer visitor or resident in foothills.

120. **CASSIN'S FINCH**, *Carpodacus cassinii*. **[sub-alp. conif.]** Sparrow size. The rose red of the head of the male sharply contrasts with its brown neck and back; the breast and rump are paler rose colored than in the Purple Finch. Female has white underparts with sharper striping than the female Purple Finch. Song like above, but breaks into a rolling, vibrant warble that ends in a *chrrr!*. When startled gives a *tay-dee-yeep!* note of alarm. Summer visitor or resident at lower elevations.

121. **GRAY-CROWNED ROSY FINCH**, *Leucosticte tephrocotis*. **[alpine]** Sparrow size. Body dark brown, touched with pink on the rump and wings; gray on sides and back of head, black on forehead. Female duller. Coarse *cheep*, *cheep* or *chee-chee-chee* calls.

123
TL 8 - 10 in.

122
TL 5½ - 6½ in.

124
TL 3¾ - 4¼ in.

125
TL 4½ - 5 in.

126
TL 6 - 6½ in.

122. RED CROSSBILL, *Loxia curvirostra.* **[sub-alp. conif.]** Sparrow size. Male is brick red with dusky wings and tail; female olive-gray with yellowish rump and underparts. Bill strongly crossed for opening pine seeds. Loud *pip-pip* cry in flight.

123. PINE GROSBEAK, *Pinicola enucleator.* **[conif. sub-alp.]** Robin size or smaller. A large, dull rose-red finch with two white wing bars on dark wings. Female gray with grayish yellow head and rump markings. Deeply notched tail. 2-3 noted whistle of *tee-tew-tew;* sharp *preer;* other twittering and whistles; rich, melodious song. From Tulare Co. to Plumas Co.

124. LESSER GOLDFINCH, *Carduelis psaltria.* **[chap. str-wd. mead.]** The male has a greenish back, black cap, black and white wings; yellow below. Female duller. Sweet, plaintive *tee-yeer* note.

125. PINE SISKIN, *Carduelis pinus.* **[conif. mead.]** Warbler size. Grayish brown above, buffy below, with dark streakings; yellow markings on wing and tail. Flies in flocks that often veer from side to side erratically, all twittering at once. Wheezy *shree-eeee* note.

(17) Family Passeridae: Weaver Finches.

126. HOUSE SPARROW, *Passer domesticus.* **[bldg.]** Male has completely black throat; gray crown; chestnut stripe behind the eye; white cheeks; brown back streaked with black; female is dusky brownish gray above and lighter below. Has a coarse, quarrelsome *cheep* or *chirp* note. Resident around buildings.

House sparrow courtship

Chickadee

Warbler

Wren-tit

Hummingbird

Woodpecker

Nuthatch

Brewer Blackbirds

Hermit Thrush

Food foraging behavior

Bird kept aloft by warm air currents rising from the earth.

Soaring

Undulating

Bird kept aloft by wing-beats.

Deep Wing-beats

Upward

Jerky

Rapid beat and glide

Straight

Erratic

Irregular rapid beat and glide

Backward

Patterns of flight

Upward Burst

REPTILES

Reptiles appear mainly during the warm months of the year, going into winter sleep in various hiding places in cold weather. Reptiles have a covering of tough scales that protect them against drying out and against enemies, whereas amphibians, such as frogs and salamanders, have a smooth to warty, moist skin. No native reptiles in California, except the rattlesnakes, are dangerously poisonous. The rest are harmless, usually do lots of good by eating excess animals and insects, and should be protected.

The reptiles listed below are divided into sections according to types of scales, head shape, etc. When identifying a reptile, study these sections to see which one it fits into. Then see what animals in the section are found in your neighborhood, and study the descriptions and pictures of these to see which ones fits your specimen. Measurements are of head and body length (i.e. from snout to vent) in lizards, frogs, and salamanders; otherwise the total length is given.

Order TESTUDINATA: Turtles and Tortoises.

Family Emydidae: Box and Water Turtles.

1. **WESTERN POND TURTLE,** *Clemmys marmorata.* **[water str-wd.]** Shell is low and smooth, olive to dark brown, often marked with a network of dark flecks and lines; underside pale yellow. Lives most of the time in water, but basks part of the time on logs and rocks in the sun; feeds on water plants, insects, and dead animals. Found at lower elevations.

1
Shell 3½ - 7½ in.

Order SQUAMATA: Lizards and Snakes.

(1) Suborder Sauria: Lizards.

A. Family Iguanidae: Iguanid Lizards. Head usually covered with large scales; belly scales comparatively small, in irregular rows; small pits or pores usually found on the underside of the thigh; skin tough; scales unequal in size.

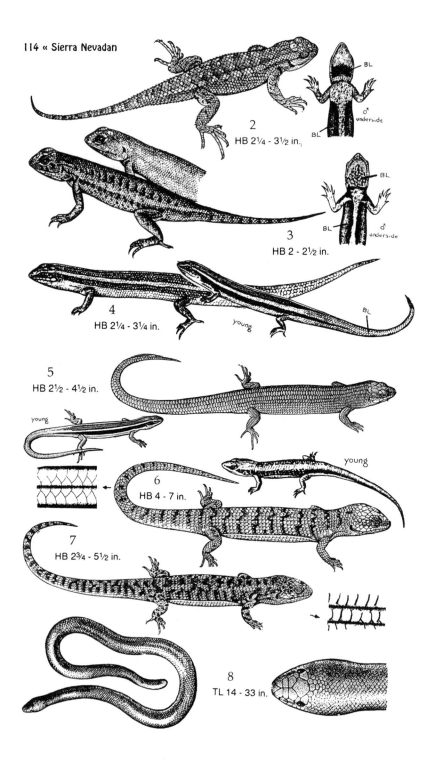

2
HB 2¼ - 3½ in.

BL

♂ underside

BL

3
HB 2 - 2½ in.

BL

BL

♂ underside

4
HB 2¼ - 3¼ in.

young

BL

5
HB 2½ - 4½ in.

young

6
HB 4 - 7 in.

young

7
HB 2¾ - 5½ in.

8
TL 14 - 33 in.

2. **WESTERN FENCE LIZARD,** *Sceloporus occidentalis.* **[chap. rocks, mead. bldg.]** Gray, brown, or blackish color with blotched pattern above; sides of belly and throat blue (sometimes lacking in female). Dorsal scales spiny, keeled, and overlapping. This species and *S. graciosus* below are spiny lizards known as "blue-bellies."

3. **SAGEBRUSH LIZARD,** *Sceloporus graciosus.* **[conif. rocks, chap. sage, sub-alp.]** Resembles the Western Fence Lizard but is smaller and has fewer scales; also unlike *S. occidentalis* the scales on back of thigh are small, smooth, granular, and not overlapping; orange or rusty colors along sides.

B. Family Scincidae: Skinks. Body scales are small, circular, and smooth.

4. **WESTERN SKINK,** *Eumeces skiltonianus.* **[conif. chap. pin-jun. mead.]** Broad brown stripe edged with black runs down the back bordered with 2 beige to whitish stripes; a dark band runs down each side. Young ones have blue tails. Comes out at dawn or dusk from hiding places under rocks, logs, or debris to hunt insects.

5. **GILBERT'S SKINK,** *Eumeces gilberti.* **[mead. conif.pin-jun.]** Adults plain olive or brown above, often with dark spotting; tail becomes brick red or orange with age; some individuals develop coppery red heads, and light lines seem to fade in larger specimens; also the scales in the light lines are bordered by an indistinct margin or the margin is absent. From Nevada Co. south, usually under rocks, debris, etc.

C. Family Anguidae: Alligator Lizards and Allies. Scales on the back keeled, feeling rough; skin fold along each side of body. Alligator lizards like a moist environment and generally find cover among dense vegetation.

6. **SOUTHERN ALLIGATOR LIZARD,** *Gerrhonotus multicarinatus.* **[conif. chap.]** Yellowish gray to reddish brown; distinct crossbands on back and tail; dark lines on belly pass down middle of scales (as shown); eyes yellow.

7. **NORTHERN ALLIGATOR LIZARD,** *Gerrhonotus coeruleus.* **[conif. str-wd.]** Olive to bluish above, usually heavily blotched or barred; dark lines on belly pass down sides of scales; eyes dark.

(2) Suborder Serpentes: Snakes. No legs; no eyelids.

A. Family Boidae: Boas and Pythons. Scales under chin not elongated, but small or rounded. Kill prey by constriction.

8. **RUBBER BOA,** *Charina bottae.* **[conif. str-wd.]** Various shades of brown to olive green above, yellowish on belly; the tail

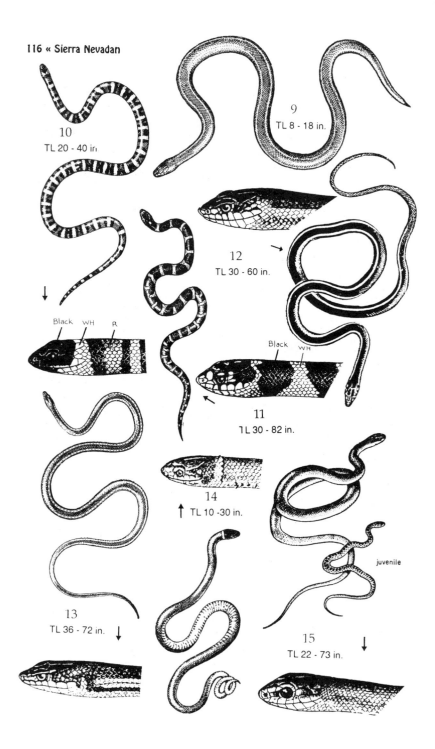

10
TL 20 - 40 in.

9
TL 8 - 18 in.

12
TL 30 - 60 in.

Black WH R

11
1L 30 - 82 in.

Black WH

14
↑ TL 10 -30 in.

13
TL 36 - 72 in. ↓

15
TL 22 - 73 in. ↓

juvenile

very blunt, looking a lot like the head; scales small and smooth; skin easily wrinkled. A secretive snake of damp ground; feeds on small mammals and birds. Docile when picked up.

B. Family Colubridae: Common Snakes. Tail tapers to a point; two or more elongated scales under the chin; belly scales are as wide as the body; head scales are usually large and symmetrical; no rattles or poison fangs.

9. **SHARP-TAILED SNAKE**, *Contia tenuis*. [**conif. str-wd. chap.**] Fairly stout body; yellowish or reddish brown or gray above; the short spine-tipped tail and alternating black and cream crossbars on belly make it easy to identify. Hides in debris and comes out when soil is damp; feeds almost entirely on slugs.

10. **CALIFORNIA MOUNTAIN KINGSNAKE**, *Lampropeltis zonata*. [**conif. mead. chap.**] A beautiful snake, banded with red, white or yellowish, and black rings; head black above; a gentle, harmless, and useful snake. Hunts lizards, snakes, small rodents, etc. A form lacking red in the colors is sometimes found around Yosemite National Park, west of the Sierra Crest.

11. **COMMON KINGSNAKE**, *Lampropeltis getulus*. [**chap. mead. str-wd.**] Banded with black or dark brown and white. Hunts birds' eggs, lizards and snakes, mainly by day; kills by constriction. Low elevation.

12. **CALIFORNIA WHIPSNAKE** or **STRIPED RACER**, *Masticophis lateralis*. [**chap.**] A swift and slender snake, black or dark brown in color with one yellow stripe on each side. A good climber in bushes; hunts with head held high off the ground. West side of the Sierra Nevada.

13. **DESERT STRIPED WHIPSNAKE**, *Masticophis taeniatus taeniatus*. [**sage pin-jun. conif.**] 15 scale rows are counted across the middle of the back. A swift, slender snake, similar in color to *M. lateralis*, but with a white stripe on each side generally bisected with a thin line of black (sometimes broken); yellowish to white on belly. Found along the east side of the Sierra Nevada at lower elevations. Climbs bushes.

14. **PACIFIC RINGNECK SNAKE**, *Diadophis punctatus amabilis*. [**conif. str-wd.**] Conspicuous yellow or orange collar against an olive to bluish gray back; head darker colored; yellow-orange to red below; belly spotted with black. It hides by day under rocks, etc.; feeds on worms, insects, and salamanders. Often curls up its bright red underside of tail to frighten enemies.

15. **WESTERN YELLOW-BELLIED RACER**, *Coluber constrictor mormon*. [**mead. chap. pin-jun.**] An agile, slender snake with

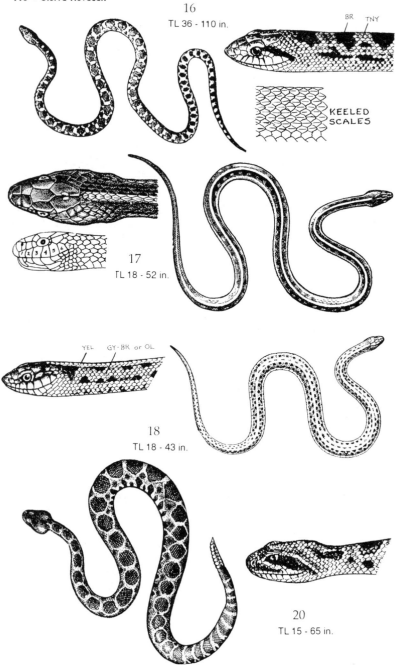

16
TL 36 - 110 in.

BR   TNY

KEELED SCALES

17
TL 18 - 52 in.

YEL   GY-BR or OL

18
TL 18 - 43 in.

20
TL 15 - 65 in.

large eyes; green, olive, or yellowish to reddish brown above; yellow on belly; young snakes are blotched like gopher snake, but scales are not keeled. This fast snake hunts with its head held high, and often catches birds.

16. **PACIFIC GOPHER SNAKE**, *Pituophis melanoleucus catenifer*. [mead.] Blotched dark brown or black on yellowish brown body; the scales on the back are keeled. Hunts rodents in their holes, also birds. Often mistaken for a rattlesnake. Lower elevations.

Garter Snakes have a distinctive pungent smell when handled; back scales are keeled. Most hunt lizards, amphibians, fish, and insects.

17. **COMMON GARTER SNAKE**, *Thamnophis sirtalis*. [mead. str-wd. water] Grayish brown in color with a distinct yellow stripe down the middle of its back as well as yellow stripes on sides; usually 7 upper lip scales (as shown); commonly red-marked between scales. In damp areas.

18. **MOUNTAIN GARTER SNAKE**, *Thamnophis elegans elegans*. [chap. conif.sub-alp.] Usually with 8 upper lip scales; general grayish brown color, with a well-defined yellow back stripe; belly pale. This terrestrial form takes cover in dense plant growth rather than retreating to water like most garter snakes.

19. **SIERRA GARTER SNAKE**, *Thamnophis couchii couchii*. [rocky streams] Inhabits same range as *T. elegans*, but frequents rocky streams with protected pools; has a faint, narrow back stripe and is yellowish or greenish gray below, often heavily marked with dark blotches. (Not illustrated.)

C. Family Viperidae, Subfamily Crotalinae. All venomous snakes are members of this subfamily, but only the rattlesnake is found in our area. Rattlesnakes have a head shaped like a blunt arrowhead, a pit behind each nostril, and rattles at end of tail.

20. **NORTHERN PACIFIC RATTLESNAKE**, *Crotalus viridis oreganus*. [chap. rocks, conif. sub-alp. str-wd.] *Poisonous*. Olive green to gray to gray-brown, with dark brown blotches bordered by light lines; color often harmonizes with ground color. Hunts small mammals by following their body heat. If you hear a rattlesnake, remain still until you have located it, and do not jump suddenly or run blindly.

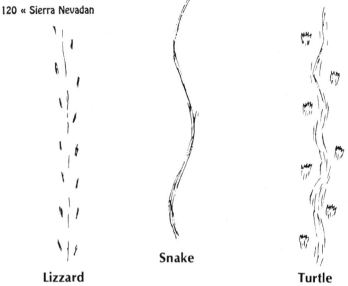

Lizzard

Snake

Turtle

Toad

Frog

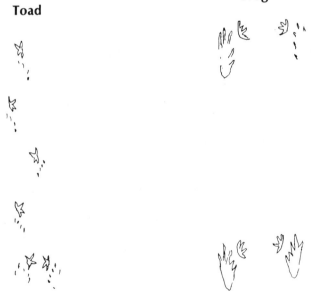

Reptile and anphibian tracks

AMPHIBIANS

Frogs, toads, and salamanders come out mainly during damp, mild weather or appear near permanent sources of water. They have smooth to warty and moist skins. None are poisonous unless eaten.

Order CAUDATA: Salamanders.

(1) Family Salamandridae: Newts.

1. **ROUGH-SKINNED NEWT,** *Taricha granulosa.* **[water conif. str-wd.]** The genus *Taricha* is characterized by a uniform dark color above and a uniform yellow, orange, or red below; also by a dark horizontal bar in the iris of the eye. *T. granulosa* is dark brown to black above, often sharply contrasting with its yellow to reddish orange undersides; light color, when present on its upper jaw, does not reach the eye; snout blunt. West slope of the Sierra from Butte Co. north and northern Coast Ranges.

2. **SIERRA NEWT,** *Taricha torosa sierrae.* **[water, str-wd. conif.]** Reddish brown to dark brown above; usually shades gradually into deep orange below; light color on its upper jaw reaches up to the eye; eyes larger than in *T. granulosa.* Found from Tulare Co. to Shasta Co. The **COAST RANGE NEWT,** *T. t. torosa,* is a similar subspecies, except for being paler orange to light yellow below.

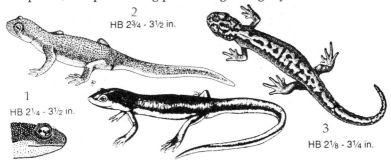

2
HB 2¾ - 3½ in.

1
HB 2¼ - 3½ in.

3
HB 2⅛ - 3¼ in.

(2) Family Ambystomidae: Mole Salamanders. Distinguished by a continuous or broken row of teeth across the roof of the mouth.

3. **SOUTHERN LONG-TOED SALAMANDER,** *Ambystoma macrodactylum sigillatum.* **[all habitats near water]** Prominent eyes; dark brown to black above with a broad, irregular, bright yellow stripe down the middle of the back, often broken into patches; dark sides and belly are white speckled; iris of eye brown flecked with gold. Found from Calaveras Co. north, but not in northern Coast Ranges. Under logs, debris.

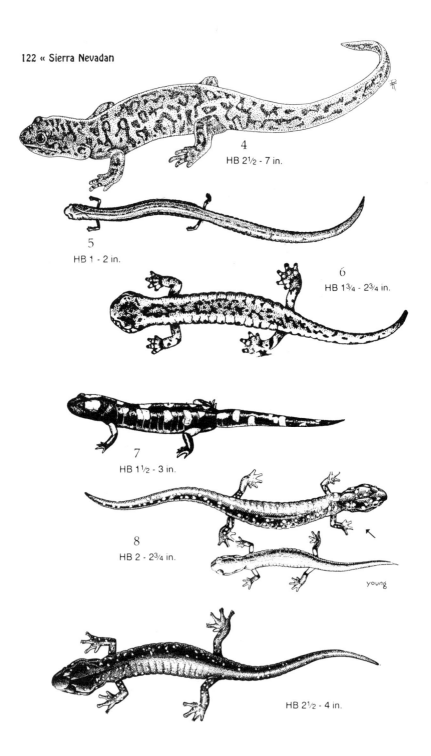

4
HB 2½ - 7 in.

5
HB 1 - 2 in.

6
HB 1¾ - 2¾ in.

7
HB 1½ - 3 in.

8
HB 2 - 2¾ in.

young

HB 2½ - 4 in.

(3) Family Dicamptodontidae: Dicamptotids.

4. **PACIFIC GIANT SALAMANDER**, *Dicamptodon ensatus*. **[water, str-wd. conif.**] Stout body with thick limbs; brown to grayish or purplish above, with black mottling; light brown to cream below, sometimes marked with dark blotches. Has barking call. Near or in cold streams, mountain lakes, or seepages. Northern Coast Ranges only.

(4) Family Plethodontidae: Lungless Salamanders. Usually smooth-skinned and spotted. Land dwellers; breath through skin.

5. **CALIFORNIA SLENDER SALAMANDER**, *Batrachoseps attenuatus*. **[str-wd. chap. conif.**] Very slender body and tiny legs; tail 1½ to 2 times HB length; dark brown to blackish body with a light back stripe; dark belly finely speckled with white. Wriggles violently on being discovered. Northern Coast Ranges and central Sierra.

6. **MOUNT LYELL SALAMANDER**, *Hydromantes platycephalus*. **[rocks, sub-alp.**] Common among granite exposures in the Sierra Nevada. Body and head flattened; granite-matching coloration; dusky below with white flecks; stubby, webbed toes. The immature form is usually blackish with green tinge. From Tulare Co. north to Sierra Co. **SHASTA SALAMANDER**, *H. shastae*. Similar to above, but body and head are less flattened and has different belly coloration. Shasta Co. only. **LIMESTONE SALAMANDER**, *H. brunus*. Similar; uniformly brown above and pale below; found in limestone areas of Mariposa County.

7. **SIERRA NEVADA SALAMANDER**, *Ensatina eschscholtzii platensis*. **[str-wd. conif.**] Tail smaller at base than immediately beyond; gray to brown above with prominent orange spots. Found from Kern Co. north; west side of the Sierra Nevada. **OREGON SALAMANDER**, *E. e. oregonensis*, a similar subspecies is found in the northern Coast Ranges, except it is plain brown or blackish above.

8. **BLACK SALAMANDER**, *Aneides flavipunctatus*. **[rocks, conif.**] Black, spotted or flecked with white or pale gold; belly dark gray to black; triangular-shaped head. Shasta and Sonoma counties north, not in Sierra Nevada mtns.

9. **ARBOREAL SALAMANDER**, *Aneides lugubris*. **[conif. brush]** Head especially wide behind prominent eyes; dark brown body spotted with yellow (spots may be weak or absent in Sierra Nevada specimens). Prehensile tail aids climbing. Calaveras to Madera County, and in Coast Ranges.

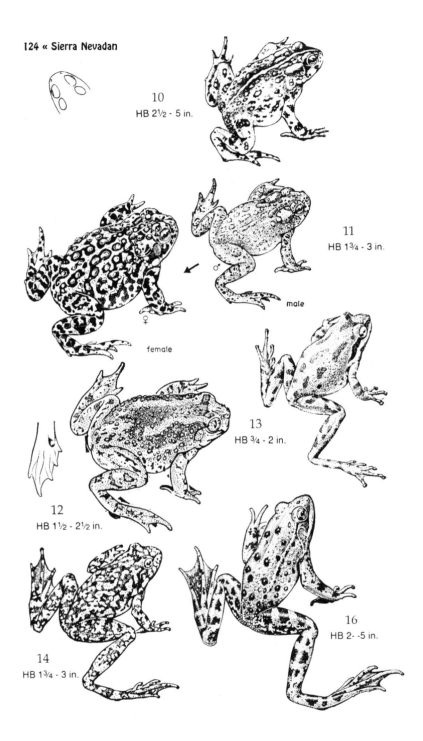

10
HB 2½ - 5 in.

11
HB 1¾ - 3 in.

♂

male

♀

female

13
HB ¾ - 2 in.

12
HB 1½ - 2½ in.

14
HB 1¾ - 3 in.

16
HB 2- -5 in.

Order SALIENTIA: Frogs and Toads.

(1) Family Bufonidae: True Toads.

10. **WESTERN TOAD,** *Bufo boreas*. **[water, mead. str-wd. conif.]** Gray to greenish with pale stripe down middle of back; warts often reddish tinged, surrounded by dark blotches.

11. **YOSEMITE TOAD,** *Bufo canorus*. **[mead. sub-alp. water]** Similar to *B. boreas*, but skin smoother and more moist. Female and young have many dark blotches contrasting to a pale background. Male uniformly pale olive to yellow-green above, with some small flecks of darker colors. Both sexes have pale throats and bellies. El Dorado to Fresno Co.

(2) Family Pelobatidae: Spadefoot Toads.
Inner edge of each hind foot has a single, black, sharp-edged "spade," which it uses to help burrow into the ground.

12. **GREAT BASIN SPADEFOOT,** *Scaphiopus intermontanus*. **[mead. chap. pin-jun. water]** Glandular hump between large, protuberant eyes; body olive to gray-green above, usually with a long, pale-colored stripe down each side. Burrows in the ground during the day, coming out at night. East side of the Sierra Crest.

(3) Family Hylidae: Treefrogs.
Toe-tips enlarged into suction discs for aid in agile climbing of rocks and trees.

13. **PACIFIC TREEFROG,** *Hyla regilla*. **[water, conif. str-wd. rocks, sub-alp.]** Color highly variable, but usually green or a shade of brown; can change its color from a dark to a light phase to match that of a leaf or bark. Black stripe from nose passes through the eyes. Active day and night; its high, two-noted *kreck-ek* is often heard.

(4) Family Ranidae: True Frogs.
No suction cups on toes.

14. **FOOTHILL YELLOW-LEGGED FROG,** *Rana boylii*. **[water str-wd. conif.]** Underside of legs and lower abdomen yellow; body mottled gray, brown, reddish, or olive above; skin granular; light colored, triangular patch below the forehead. Found up to about 6,000 ft.

15. **MOUNTAIN YELLOW-LEGGED FROG,** *Rana muscosa*. **[water, mead. sub-alp. alp]** Similar to *R. boylii*, but more yellowish on belly, and brownish with dark spots or lichen-like markings above. 4,500 to over 12,000 ft. Not in Coast Ranges. (Not illustrated.)

16. **RED-LEGGED FROG,** *Rana aurora*. **[water, str-wd. mead.]** Chiefly a pond frog. Red on lower abdomen and underside of hind legs; reddish brown to gray and dark mottled above, usually

with a dark mask over the eyes bordered by a light jaw stripe. More common at lower altitudes; west side of the Sierra.

17. **NORTHERN LEOPARD FROG**, *Rana pipiens*. **[water, mead. sub-alp.]** Brownish or green above, with large round to oval, pale-bordered dark spots; light stripe along upper jaw; yellowish white on belly. Found at Lake Tahoe and a few other high mt. lakes to the north.

18. **BULLFROG**, *Rana catesbeiana*. **[water]** Very large greenish to brown frog; whitish mottled with gray below. Eardrums as large as the eyes. Eats large insects, other frogs, and mice. The famous, deep bass *jug-o-rum!* call is heard from the male at night. Introduced.

17
HB 2 - 4½ in.

18
HB 3½ - 8 in.

FISH

Fish caught in sport fishing are found in the Sierra Nevadan region in the clear mountain streams and lakes, but there are also many other interesting fish to be seen in these waters and some are shown here. Before looking at the descriptions of the different fish, study the drawing of a generalized fish below to become familiar with the names of the different parts of the body. Sometimes you may catch or see a rare or introduced fish not described here. Because so many kinds of fish have been brought in from outside the state or transplanted within the state, it can become rather confusing. For this reason, specific ranges for the fish are not given in every case, and fish that have been introduced are marked with an asterisk(*). Lengths given are for the average size of adult fish, though specimens much larger are also found, because fish continue to grow as long as they live.

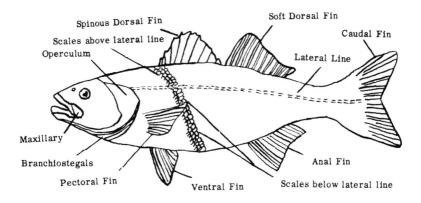

Parts of a fish

A. Family Salmonidae: Salmon and Trout. These fish are characterized by having an adipose (or fleshlike) fin on the rear of the back, while the main dorsal fin is shorter than the head and has less than 15 rays. Ventral fins are always on the middle of the body, never just below the pectoral fin. Swift game fish.

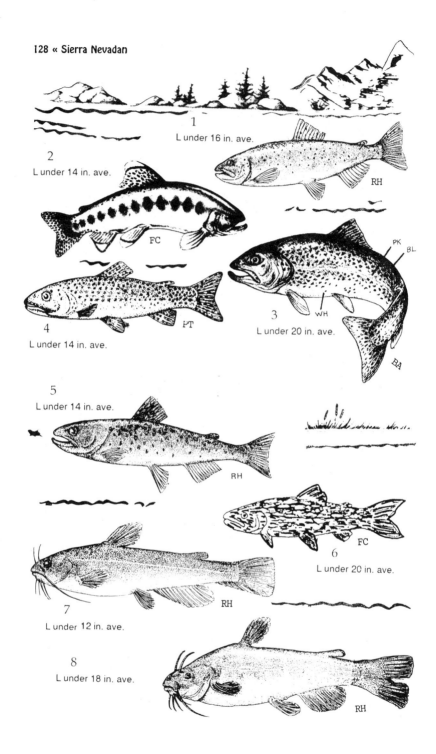

1
L under 16 in. ave.

RH

2
L under 14 in. ave.

FC

4
L under 14 in. ave.

PT

3
L under 20 in. ave.

PK
BL
WH
BA

5
L under 14 in. ave.

RH

6
L under 20 in. ave.

FC

7
L under 12 in. ave.

RH

8
L under 18 in. ave.

RH

1.* **BROWN TROUT**, *Salmo trutta*. Dorsal fin has black spots; red and dark brown spots on head and body in varying degrees, some spots surrounded by pale halos; olive to dark brown above, silvery yellow-brown sides, whitish to yellowish below.

2. **GOLDEN TROUT**, *Oncorhynchus aquabonita*. The most beautiful trout of the Sierra; golden yellow lower sides, red-orange to bright red cheeks, belly, and side stripes; dark olive upper back; red side stripe marked with a series of dark parr marks; lower fins orange. Native to the headwaters of the Kern River, but introduced into other mountain lakes and streams in the Sierra Nevada.

3. **RAINBOW TROUT** or **STEELHEAD**, *Oncorhynchus mykiss*. Small black spots on back and most fins. Olive green, light brown, or steel blue above; rose band along middle of sides; silvery to pale below. Note: Sea-running rainbow trout are called steelhead, while if they remain in fresh water they are called rainbow.

4.* **CUTTHROAT TROUT**, *Oncorhynchus clarki*. Varies widely in color and pattern, but is distinguished by the red or pink "cutthroat" streak on each side of the lower jaw; dorsal rays 9-11, generally 10; body and fins covered with many dark spots. Interbreeds with golden and rainbow trouts.

5.* **BROOK TROUT**, *Salvelinus fontinalis*. Back mottled olive to black; front edges of red lower fins white; scales very tiny; light spots on dark background; usually pink or red spots on sides. Mainly in streams.

6. **LAKE TROUT**, *Salvelinus namaycush*. May reach a length of 4' and weigh over 100 pounds. A variable gray to dark olive color, colored with numerous light spots; tail fin deeply forked; belly silvery, sometimes lightly spotted. Lakes, especially Tahoe.

B. Family Ictaluridae: Bullhead Catfishes. Body without scales; conspicuous barbels or "whiskers" around the mouth.

7.* **BROWN BULLHEAD**, *Ameiurus nebulosus*. Dark brown or gray on back gradually shading to light yellowish brown on sides; brown to black mottling or spots on body; fins brown; caudal fin squared; strong saw-toothed spine on pectoral fin. Likes pools or sluggish water in rivers, lakes, and ponds. **BLACK BULLHEAD,** *Ameiurus melas*. Very similar, but darker; its pectoral spine is more weakly barbed and it has no mottling on body.

8.* **WHITE CATFISH**, *Ameiurus catus*. Upper body and top of head gray to blue-black; sides and belly white to light yellow; caudal fin moderately forked; anal fin with 19-23 rays.

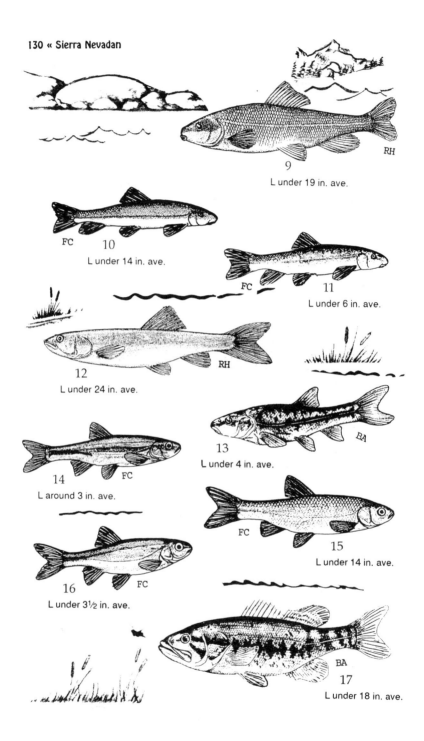

RH

9

L under 19 in. ave.

FC 10

L under 14 in. ave.

FC 11

L under 6 in. ave.

12

RH

L under 24 in. ave.

13

BA

L under 4 in. ave.

14 FC

L around 3 in. ave.

15

FC

L under 14 in. ave.

16 FC

L under 3½ in. ave.

17

BA

L under 18 in. ave.

C. Family Catostomidae: Suckers. Down-pointing, sucker mouth with large lips. Generally slow plant feeders along the bottom.

9. **SACRAMENTO SUCKER,** *Catostomus occidentalis.* Olive to gray or brown above, white to dirty yellow-gold below; 56 to 75 lateral line scales; 12-15 dorsal rays. Most common in clear, cool pools of streams.

10. **TAHOE SUCKER,** *Catostomus tahoensis.* Lateral line scales 82-95. Dark olive above contrasts to white or yellow below. Found in Lake Tahoe and in the streams entering the Great Basin; introduced in the Feather and upper Sacramento rivers, etc.

11. **MOUNTAIN SUCKER,** *Catostomus platyrhynchus.* Dusky gray to olive above, sometimes with a dark stripe along the side; white to yellow below. Breeding male with bright red stripe on side. Found in streams entering the Great Basin, in the north fork of the Feather River, and the upper Sacramento River system.

D. Family Cyprinidae: Carps and Minnows. One dorsal fin, jaws without teeth, and cycloid scales.

12. **SACRAMENTO SQUAWFISH,** *Ptychocheilus grandis.* Back dark olive brown; sides silvery grayish; no barbels; 8 dorsal fin rays. Clear, warm streams and rivers at lower elevations.

13. **SPECKLED DACE,** *Rhinichthys osculus.* Usually dark olive above heavily speckled with black; some gold specks; dark stripe runs along side through eye to the snout; barbel at the end of the maxillary. Widespread on both sides of the Sierra Nevada.

14. **LAHONTAN REDSIDE,** *Richardsonius egregius.* Olive-gray to brown above; dark midlateral stripe bordered above by clear to yellow streak; red above base of pectoral fin; breeding male shows bright red along lower sides. Found in streams that flow into the Great Basin; also in upper Sacramento River system.

15. **TUI CHUB,** *Gila bicolor.* Dark olive to brassy above; sides lighter; small mouth does not reach front of eye; young fish has dark stripe along side; large adult may have yellow to copper fins with pink, red, or orange bases. Quiet water with vegetation.

16. **CALIFORNIA ROACH,** *Hesperoleucus symmetricus.* Dusky gray to steel blue above; dull silvery sides with dark lateral stripe. Breeding male has red-orange on chin and gill cover. Front of dorsal fin is far behind front of pelvic fins; has 32-38 scales in front of dorsal fin. In Sacramento and San Joaquin drainages.

E. Family Centrarchidae: Basses and Sunfishes.

17.* **LARGEMOUTH BASS,** *Micropterus salmoides.* Brownish green above, only slightly lighter on sides; belly white or creamy;

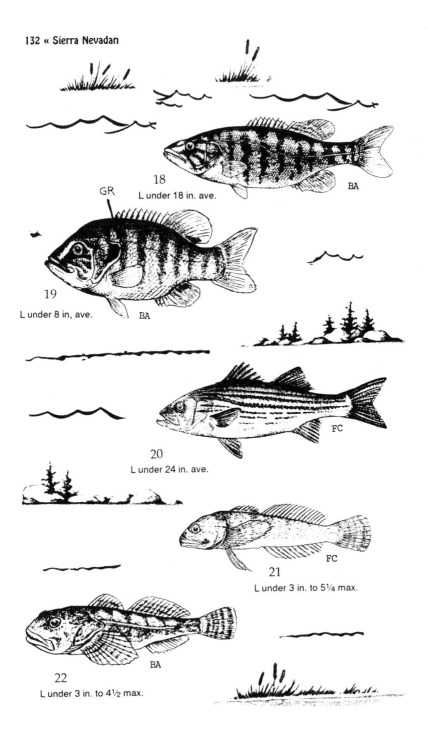

18
L under 18 in. ave.

BA

GR

19
L under 8 in, ave.

BA

20
L under 24 in. ave.

FC

21
L under 3 in. to 5¼ max.

FC

22
L under 3 in. to 4½ max.

BA

broad black stripe (often broken into a series of blotches) runs along side to the eye; very large mouth with upper jaw reaching beyond the eye; dorsal fins united by a membrane. Found at lower elevations in vegetated water of west slope rivers, ponds, and creeks. Common in impoundments.

18.* **SMALLMOUTH BASS,** *Micropterus dolomieu.* Dark brown to dark olive above; irregular dark bars or mottling on yellow-green sides; red eyes; mouth small and not reaching beyond the eye as in the Largemouth; also has no deep notch in its dorsal fin. In west slope rivers with clear, gravel-bottom runs and flowing pools.

19.* **GREEN SUNFISH,** *Lepomis cyanellus.* Robust fish; yellowish olive color, sides sometimes with dusky bars; belly pale olive; adult has a large black spot at rear of second dorsal and anal fin bases; edges on second dorsal, caudal, and anal fins yellow or orange. In quiet water of west slope rivers and ponds.

Crappies (with wavelike marks) and Bluegills (orange throat) are also occasionally found in quiet waters with vegetation. All are good to eat.

F. Family Moronidae: Temperate Basses.

20.* **STRIPED BASS,** *Morone saxatilis.* Body elongate; back dark olive to blue-gray; silvery white sides have 6-9 dark uninterrupted stripes; white belly; two separate dorsal fins. This is a marine fish, but it travels rivers far upstream to spawn. Can reach a length of 6½ ft.

G. Family Cottidae: Sculpins.

21. **PAIUTE SCULPIN,** *Cottus beldingi.* No prickles. Brown to grayish and mottled in color; dorsal fins separate to base; one large spine present on the front edge of the gills, no other smaller spines below; orange edge on first dorsal fin. Found in streams draining into the Great Basin, abundant in Lake Tahoe.

22. **RIFFLE SCULPIN,** *Cottus gulosus.* Prickles confined to area behind pectoral fin base. Brown to grayish olive above with dark brown mottling; white to yellow below; dorsal fins joined at base. Found in most of Sierra Nevadan range in sand and gravel riffles of creeks and rivers.

SUGGESTED REFERENCES

Plants

Abrams, Leroy. *Illustrated Flora of the Pacific States*, 4 volumes. Stanford: Stanford University Press, 1960.

Munz, Phillip, and David Keck. *A California Flora*. Berkeley: University of California Press, 1963.

Weeden, Norman F. *A Sierra Nevada Flora*, 3rd Ed. Berkeley: Wilderness Press, 1986.

Mammals

Burt, William H., and Richard P. Grossenheider. *A Field Guide to the Mammals: Fieldmarks of all North American species found north of Mexico*, 3rd Ed. Boston: Houghton Mifflin Co., 1976.

California's Wildlife, Vol. 3, Mammals. Sacramento, CA: Department of Fish & Game, 1990.

Ingles, Lloyd G. *Mammals of the Pacific States: California, Oregon, and Washington*. Stanford: Stanford University Press, 1965.

Jameson Jr., E. W., and Hans J. Peeters. *California Mammals*. Berkeley: University of California Press, 1988.

Birds

Beedy, Edward C. and Stephen L. Granholm. *Discovering Sierra Birds: Western Slope*. Yosemite Natural History Association, 1985.

Brown, Vinson, Henry Weston Jr., and Jerry Buzzell. *Handbook of California Birds*. Happy Camp, CA: Naturegraph, 1986.

Peterson, Roger Tory. *Western Birds*. Boston: Houghton Mifflin, 1990.

Udvardy, Miklos D.F. *The Audubon Society Field Guide to North American Birds: Western Region*. New York: Alfred A. Knopf, 1977.

Reptiles and Amphibians

Behler, John L. and F. Wayne King. *The Audubon Society Field Guide to North American Reptiles and Amphibians*. New York: Alfred A. Knopf, 1979.

Brown, Vinson. *Reptiles and Amphibians of the West*. Happy Camp, CA: 1974.

Savage, Jay. *An Illustrated Key to Lizards, Snakes and Turtles of the West*. Happy Camp, CA: Naturegraph, 1989.

Stebbins, Robert C. *A Field Guide to Western Reptiles and Amphibians*, 2nd Rev. Ed. Boston: Houghton Mifflin Co., 1985.

Fish

Lee, David S., et. al. *Atlas of North American Freshwater Fishes*, North Carolina State Museum of Natural History, 1980.

Page, Lawrence M. and Brooks M. Burr. *A Field Guide to Freshwater Fishes, North America North of Mexico*. Boston: Houghton Mifflin, 1991.

Thompson, Peter. *Thompson's Guide to Freshwater Fishes*. Boston: Houghton Mifflin, 1985.

Williams, James D. et. al. *The Audubon Society Field Guide to North American Fishes, Whales, and Dolphins*. New York: Alfred A. Knopf, 1983.

INDEX